Silk Sails

CALVIN D. EVANS

SILK SAILS

Women of Newfoundland and Their Ships

BREAKWATER BOOKS LTD.
JESPERSON PUBLISHING • BREAKWATER DISTRIBUTORS

100 Water Street • P.O. Box 2188 • St. John's • NL • A1C 6E6
www.breakwaterbooks.com www.jespersonpublishing.ca

Library and Archives Canada Cataloguing in Publication
Evans, Calvin D.
Silk sails : women of Newfoundland and their ships / Calvin D. Evans.

Includes bibliographical references and index.

ISBN 978-1-55081-242-8

1. Women in fisheries--Newfoundland and Labrador--History.
2. Women fishers--Newfoundland and Labrador--History. 3. Fisheries--Newfoundland and Labrador--History. 4. Businesswomen--Newfoundland and Labrador-- History. I. Title.

SH224.N7E93 2008 338.3'72708209718 C2008-905007-XISBN

The Canada Council for the Arts | Le Conseil des Arts du Canada

We acknowledge the financial support of The Canada Council for the Arts for our publishing activities.

We acknowledge the support of the Department of Tourism, Culture and Recreation for our publishing activities.

Canada

We acknowledge the financial support of the Government of Canada through the Book Publishing Industry Development Program (BPIDP) for our publishing activities.

Printed in Canada.

ENVIRONMENTAL CHOICE • CHOIX ENVIRONNEMENTAL

"Let our sons in their
youth be as grown-up plants, and
let our daughters be as corner pillars
fashioned as for a palace."

Psalm 144:12

This book is dedicated to all
women of the sea in general,
and lovingly to two women in particular:

my mother,
Mary Jane (Lidstone) Evans,
who died in 1949 at age 42

and my mother-in-law,
Minnie Goldie (Waterman) Locke,
who died in 2002 at age 96.

CONTENTS

FOREWORD

Changes in the status of women during modern times – the right to own property, to vote and participate in public life, and to choose their own careers – tend to obscure achievements of individual women, and women generally, in the past. This has certainly been the case in Newfoundland and Labrador.

Calvin Evans performs a most valuable service by researching the historical role of women as owners of ships and fishing properties (rooms) in Newfoundland. Additionally this book is a significant contribution to the recent literature focusing on the role of women in spheres of life traditionally reserved for men.

Only recently has Newfoundland history, mostly written by men, begun to acknowledge the fundamental contributions of women in the settlement of Newfoundland and in traditional basic economic activities such as the fishery and shipping. Hilda Chaulk Murray's seminal work *More than Fifty Percent: Woman's Life in a Newfoundland Outport, 1900-1950* (1979) helps redress this neglect by showing that outport women were fully involved in all aspects of earning a livelihood in the pre-Confederation era. Calvin Evans now illustrates how women participated and shared more fully in other "big events" of Newfoundland life than previously known or perceived even by scholars of maritime history. He also identifies these women personally and in the context of their families and communities.

It could be argued that the arrival and continuing presence of women in the resident population played a crucial role in the successful establishment of early communities in Newfoundland. Among other influences, determined and resourceful women were collectively a major force, perhaps the most important, in ultimately helping undermine formal attempts to prevent and restrict the growth of a permanent population.

Historically, seafaring and fishing were activities from which women were normally excluded because of gender proscriptions, often codified into legal systems, related to property ownership and inheritance. It was traditionally held, for example, that women, the so-called weaker sex, were not physically or emotionally capable of dealing with the challenges of seafaring or coastal and deep-sea fishing, and therefore had no place on ships, in fishing boats, or at sea. Despite these norms and English Common Law generally (giving husbands almost unlimited rights over property belonging to their wives), there were exceptions and circumstances in which women did assume and perform male roles. It is these exceptions in the context of the Newfoundland fishery that Calvin Evans identifies, analyzes and describes.

Traditionally women had the acknowledged right to become owners or shareholders of ships when their husbands died without male heirs of sufficient maturity or suitability to assume these roles. Thus in eighteenth and nineteenth century England, female owners were most frequently widows who became temporary shareholders until any sons they might have were of age and experience to replace their fathers. In rarer cases shares were held by "spinsters." Calvin Evans documents a much more complex pattern of female shipowners in Newfoundland. Although most were "widows" and "spinsters," many were actively involved in fishing and shipping enterprises. Indeed, he found women in sundry roles as owners – sole owners, joint owners, co-partners in trade (with other women or men), and managing owners – and discovered it not at all uncommon for women to purchase shares as well as to inherit them.

Silk Sails explores thoroughly, from ship registers, court records, wills and deeds, and field research (interviews), instances in which Newfoundland women from at least the time of Lady Sara Kirke of Ferryland in the 1650s formed part of the owners (and operators) of trading and fishing vessels and fishing enterprises. In modern times women have become even more prominent in these roles. In 1949, for example, upon the death of her husband, Marie Smart Penny of Little Bay assumed the presidency of John Penny & Sons of Ramea and directed the operation of a major fish-processing company. In 1967 she was elected president of the Fisheries Council of Canada, thus becoming the first woman to hold that position. But there were many more women like Marie Penny, not as prominent, but nevertheless influential in their own respective domains.

In all, *Silk Sails* profiles more than 500 exceptional women who influenced the development, growth and maintenance of their respective fishing and seafaring communities, especially in Conception Bay, the Burin Peninsula, and along the South Coast. Their stories are well worth the telling and Calvin Evans does a splendid job of recounting them.

W. Gordon Handcock
Professor Emeritas, Memorial University of Newfoundland
March 2008

INTRODUCTION

This book is about a piece of lost history: the knowledge that women played important roles in early fishing enterprises and made significant contributions to the early economies of the Atlantic region. The information presented in *Silk Sails: Women of Newfoundland and Their Ships* documents that women along the Atlantic coast were boat-owners in charge of fishing ventures for at least 300 years (since about 1650 and possibly earlier) and that women have been shipowners involved in both the local fishery and foreign-going business for at least 200 years (since about 1800).

The problems encountered and conquered in the early maritime region produced hardy, durable, tenacious and feisty women, who have received far less attention in the historical records than they rightly deserve. Most history has been written by men, and, until the recent advent of scholarly research in women's history, men have uniformly neglected to give women their rightful due. Perhaps this neglect was largely unintentional since women were usually little involved in the "big events," and their essential roles in the background seldom played through into the foreground in obvious ways. The irony is that the public man so prominently depicted in history is often largely the result of being protected and sheltered from inner concerns while being championed and fitted for public affairs *by women* (mothers). Mothers tend to protect their sons from the inner struggles; daughters are expected and conditioned by mothers to endure the same stress that the mother has endured and to carry the same weight and the same burdens. Daughters are trained for the harder, tougher mental and emotional tasks. John Langdon-Davies suggested in 1927 in the book *A Short History of Women* that women possess "a greater singleness of purpose and a greater fund of imagination" than men, and in this dual possession is their unique strength. While women shamelessly mother their sons into privileged softies, they mother their daughters into enduring, strong women, modeled after themselves.

Because of the lack, even non-existence, of a large body of documentation supplementary to the ship registers for Newfoundland and the Maritime provinces, it is difficult to substantiate women's roles in early fishing economies. Court records, plantation books, occasional references in the literature, and oral sources all help towards this end. I used all these sources extensively in my research, which set out to show that: i) women were involved in a substantial way in the early economy and society of Newfoundland and that they were owners of boats, ships and waterfront properties associated with the fishery; ii) it was real ownership involving real money; iii) although it was real ownership, society has collectively forgotten that there was a time in maritime history when women's substantial roles were being played out; iv) ownership was a natural activity and evolved out of necessity and opportunity; v) women possessed the requisite skills to be real partners in business ventures; and vi) older women modeled shipowning and property ownership for younger women.

The women growing up along the Atlantic coast were durable women fitted to carry the world on their shoulders – forget about Atlas! They accomplished extraordinary things such as running businesses and owning their own ships in a period when, according to the accepted storyline, women were thought to have been "barefoot, pregnant, and in the kitchen." The documentation presented in *Silk Sails* demonstrates conclusively that discarding the near-myth of the passive woman in early society is long overdue.

CHAPTER ONE

Women and the Sea

It is a serious error to think of the sea as the exclusive domain of men. Indeed, men and women have shared work on the sea for many generations and in some most unusual partnerships.

From earliest times, women have functioned on the sea as pirates, warriors, traders, whalers, workers, travelers, navigators, captains, and as working and supportive wives. They have even functioned at the edge of the sea as "fish-makers" and shipbuilders.

Why are we surprised by this? Both because there has been a concerted effort to write history from a man's point of view and a tendency to ignore women's involvement, as if women were invisible or unworthy of mention, and because the writing of women's history in a structured form is a relatively recent phenomenon. Attempts to depict the sea as belonging to men and the shore as belonging to women, and to depict the sailor and seaman as hard, tough and anti-domestic, appear to be myths perpetrated by shipowners and investors. It may well have served an exploitative purpose such as bolstering the mystique that helped attract the very best men to a kind of exclusive club and a select inner circle.

The averred connection between women, ships and bad luck seems also to have been a fiction promoted by shipowners who wanted more consistent hard work from their male crews, without the added distraction of having women on board. The presence of women might have obliged men to act more decently and with less cruelty and would have required a curb on their infamous shipboard profanity. How the fiction of bad luck arose is uncertain. It seems not to have originated with sailors of European or American descent. Mary Chipman Lawrence, who sailed with her husband Samuel on board the whaler *Addison* from 1856 to 1860, wrote in her journal about the natives of the Marquesas Islands: "The women are not allowed to go in a canoe; it is 'taboo' to them. If they wish to go to another bay, they are compelled to go over rocks and ridges while their husbands go in a canoe. They may go in a whaleboat, however, when opportunity offers." This may have been an interpretation on Mary Lawrence's part, or the taboo may have been only occasional and associated perhaps with a woman's menstrual period. Apart from such rare instances we may assume that women were associated with the sea in the same natural relationship as men. Only their physical differences would have affected some of the roles they played.

There was a recognition in ancient Greece of a woman's right to own a ship, but when she wished to sell or dispose of it she was required to request a man's help if not also his permission. This requirement would likely have applied to a family member only.

Even the titles of books have assumed a natural relationship between men and the sea: *Wooden Ships and Iron Men*; *Men, Ships and the Sea*; *Atlantic Conquest: The Men and Ships of the Glorious Age of Steam*; and *Men and Ships in the Canada Trade.* Actually, there were many women involved in the Canada Trade, and they will be included in a subsequent work about Quebec and the Maritime provinces.

Men have always attempted to carry the mystic female presence to sea with them in the guise of naming their ships after women, having their ship christened by a woman, wearing tatoos of women, or designing a ship figurehead in the female form.

Women, as well as men, have written about the sea from earliest times, and women's diaries, journals and articles are being recovered and published to demonstrate their unique insights and observations and to form an essential component of women's history. Unfortunately, there are not enough of these, and most of them deal with sea voyages rather than life at the edge of the sea.

One of the very best books dealing with women's historic relationship with the sea is Linda Grant DePauw's *Seafaring Women*. The author demonstrates the amazing breadth of women's involvement with the sea as she writes about women as pirates, warriors, whalers and traders. I am indebted to her book for much of the information in this chapter, especially for the stories of Hannah Burgess and Mary Patten. Many additional sources have been consulted as well.

Perhaps the earliest known female pirate was Alvida, who operated in the North Atlantic from a Scandinavian base and whose exploits were recorded by Saxo Grammaticus, a twelfth-century historian. Another was Lady Killigrew of Cornwall, wife of Sir John Killigrew, and a contemporary of Queen Elizabeth I. Grace O'Malley, an Irish pirate, operated during the same period. With her father's support, Grace supplanted her brother as heir and commanded her family's fleets and castles after her father's death. Mary Lindsey of Plymouth, England, married a pirate, Eric Cobham, and became a pirate captain herself. They operated pirate ships out of Poole and Plymouth.

Women also served on ships as warriors or as crew members on fighting ships. During the eighteenth century, Anne Bonny and Mary Read were crew members aboard a pirate ship. During the American Revolution, Fanny Campbell of Lynn, Massachusetts, captained a privately owned ship that fought against the British. About the same time, a woman was in charge of the French privateer *La Bougourt* in the West Indies where she attacked British ships. Throughout the eighteenth and nineteenth centuries, women either worked voluntarily as crew members of fighting ships (often disguised as men) or were pressed into service during crises aboard ship; in the latter case they were usually paid for their service.

In 1730 Dame Suzanne, nee LeGros, widow of Nicholas le Pelley, purchased the Island of Sark from Sir Charles de Carteret. Dame Suzanne had been born on the Island of Sark and the le Pelley family were from Guernsey. About 1755 the widow of Daniel le Pelley, "Dame of Sark," Dame Elizabeth Etienne, who had already served a term in an ecclesiastical position as "douzenier," which would have been extremely rare for a woman in those days, dismissed a priest for misconduct and barred the door of the church. The controversy had to do with Dame Elizabeth's decision to award "livings" (clergy, clerical wages, or lands) to lapsed Catholics. The Anglican Church disagreed. Though the Dean of Guernsey challenged Dame Elizabeth's action initially, he relented in the end and the matter was resolved the following year. Sark remained a possession of the le Pelley family until 1852. Strong women such as these may have been rare, but there were women of such stature here and there, now and then. Dame Elizabeth had, in fact, taken over commercial operations in the Jersey Islands after her husband's death. Because of the connection between these islands and Newfoundland, it is possible that civil law governing property may have been at least temporarily established in Newfoundland.

In much more recent days, there is the story of Katherine Dorey taking over the Guernsey family firm of Dorey Shipping Co. Ltd. when her husband Peter drowned in 1979. The firm had been in business for about 100 years as a tramp shipping company carrying cargoes of coal, wheat, and stone, as well as other commodities. In their shipyard at St. Samson they built several small vessels and repaired others, and Peter set up the Condor hydrofoil service sometime in the 1970s. Katherine took over the management of this firm at a very difficult time when British shipping was in a depressed condition. Condor was sold in 1984 and the Doreys' other ships eventually came under the management of the Fishers of Barrow.

American women in an earlier period were involved in the whaling trade, and the names of Martha's Vineyard, New Bedford, Salem and Nantucket figure prominently in the whaling business.

Martha Smith owned a fleet of whaling ships by 1718. Kezia Coffin Fanning of Nantucket operated a large shipping business during the American Revolution while her husband, John Fanning, was away on whaling voyages. Kezia was loyal to the British and enjoyed their protection during the war as she consolidated her trade and took mortgages on wharves, warehouses and other assets of her customers. By the end of the war she held mortgages on much of the valuable property of Nantucket and owned a townhouse on Center Street and a sumptuous country place. When the war ended with a win by the revolutionaries, Kezia fled to Halifax where she was later imprisoned. She eventually lost all her property and her husband died on the island.

Something of the quality of the women of this period may be seen in Mary English, wife of Philip English, who was the major shipowner of Salem during the period 1680 to 1750. She had the best education available at that time and wrote very well. She was also a devout member of the Established Church. In the witchcraft frenzy, Mary English was "cried against," arrested and imprisoned. The manner of her arrest is worth noting. Guards came to her bedroom to take her away, but she refused to move until morning. After morning devotions she tended to the needs of her family, outlined plans for her children's education, kissed them goodbye, and told the officer in charge that she was ready to die. After she had been in prison for six weeks, her husband, who visited her regularly, was also arrested. They were both transported to Boston where they later escaped from jail and were taken to a safe place in New York.

The breakaway from European culture and education in the early American colonies caused a diffusion of roles for both men and women. Daniel Boorstin writes: "Although our knowledge is only fragmentary, evidence suggests that women in colonial America were more versatile, more active, more prominent, and on the whole more successful in activities outside the kitchen than were their English counterparts." Wives and daughters took advantage of learning opportunities provided through the husband's and father's system of household manufacturing, and women became involved in printing and publishing; they became merchants and tradespersons and often acted as medical practitioners. A woman's cooperation and energy were essential for many a family business to succeed. In this environment mothers became responsible for education in the family, many of them having been educated by their own parents who saw the need for studies in various foreign languages and "the graver sciences." Boorstin continues: "Even such fragmentary evidence suggests that women in the colonies were successful in more different activities and were more prominent in professional and public life than they would be again until the 20th century." The law protected women in the colonies well beyond what the British common law provided, and married women enjoyed unprecedented rights in their ability to initiate new businesses and continue these. The stories of women's bravery and courage and fighting spirit even in colonial warfare illustrate their ability to rise to new challenges and to be undaunted in the face of the worst atrocities.

Hector St. John de Crevecoeur published his *Letters to an American Farmer* in 1782 after a visit to Nantucket, during which he was very impressed with the whaling wives of the island. Not only were they industrious in the home and in social affairs, they were astute business heads as well. He writes: "As the sea excursions are often very long, their wives in their absence are necessarily obliged to transact business, to settle accounts, and in short, to rule and provide for their families. These circumstances being often repeated, give women the abilities as well as the taste for that kind of superintendency, to which by their prudence and good management, they seem to be in general very equal. This employment ripens their judgement, and justly entitles them to a rank superior to that of other wives, and this is the principal reason why those of Nantucket as well as those of Montreal are so fond of society, so affable, and so conversant with the affairs of the world." What made the wives of Montreal similar to the wives of Nantucket is that their husbands, as merchants and traders, were away for similarly long periods of time, trading with the native people. Crevecoeur refers to the ingenious Aunt Kesiah and identifies her husband, the richest person now in the island, as Mr. C- - - -n. It is obviously the Kezia Coffin Fanning referred to earlier, so the error must be in the husband's name. It is a well known and recorded fact, says Crevecoeur, that Kesiah is the secret of her husband's success, for she started trading at first with pins and needles and kept a school. She then moved into trading with "more considerable articles" and "laid the foundation of a system of business, that she has ever since prosecuted with equal dexterity and success." Her business connections extended even to London, England.

Until at least 1880, except for the very earliest period, women on Nantucket Island outnumbered men by at least four to one. Many of these were single women or widows, or married women separated from their husbands for the duration of a whaling voyage, which usually lasted three or four years. These conditions produced a practical, forceful woman who did not shrink from responsibility or intellectual accomplishment. In addition to their accomplishments on land, Nantucket women were among the first to accompany their husbands to sea on the long whaling voyages. In 1817, Mary Russell of Nantucket went to sea with her husband Laban aboard the whaler *Hydra*, and thus became the first woman to establish a home on a whaling ship. Within a few years many would follow her example.

Many of the women who went to sea with their husbands on whaling voyages were, fortunately, tireless diarists, and we therefore have an excellent record of their unique perspectives, the difficulties encountered, the joys of life aboard ship, and a relatively normal life where children were born, raised, educated, and taught many of life's most practical lessons. The presence of women on board ship had a civilizing influence on both captain and crew, as both captains' and crew members' diaries attest. Linda Grant DePauw observes: "For the Victorians, homemaking did not mean doing domestic chores so much as maintaining an emotional environment of peace and moral uplift. Men, it was believed, required the presence of a 'mother' to keep them well behaved and civilized."

But there was another factor at work in the man-centered world of the Victorian era (ironic as that statement may seem). It is reflected in Julia C. Bonham's comment in a 1977 article in the magazine *American Neptune*: "But the men who treated women as their intellectual and physical underlings were quite willing to regard the ladies as their moral and spiritual superiors." Not only did women bring a civilizing influence to life on the ship, they contributed to a balanced social life aboard that generally made difficult situations more bearable.

Mary Chipman Lawrence's journal of 1856 to 1860 has been published under the title *The Captain's Best Mate,* which is, apparently, the title that her husband attributed to her. A picture emerges in her journal of a quite normal life on board ship, a life that is filled with writing, painting, music and song, teaching, "gamming" (visiting with occupants of other ships), dispensing medicines, letters from home, danger and even death. The voyage covers seven cruises from New Bedford to the Pacific Ocean, to the Arctic Ocean, to New Zealand, and home to New Bedford. This journal is typical of many others, not as reflecting an imposition made upon women, but as a task undertaken voluntarily and willingly, and without blaming either persons or circumstances.

Eliza Azelia Williams accompanied her husband Thomas William Williams aboard the whaler *Florida* from 1858 to 1861. During the voyage, Eliza had two children and kept a detailed diary of the trips to the Indian and Arctic oceans. One of the children born on this first voyage, William Fish Williams, paid a tribute in later years to his mother in the words: "A woman can be depended on to show true nerve and grit at the crucial moment better than a man" (DePauw).

Some of the most amazing stories about women and the sea have to do with women on trading ships. The story of Rosina Annie Slade, 1865 to 1964, is a fine example. Annie was the daughter of Richard Harding of Appledore, England. Her father was master of the trading smack *Dahlia*. His two older sons had gone off to sea on the square-riggers and he had only a very young son remaining, so Annie, having just finished school, decided to go on the ship with her father. She soon mastered the compass and how to steer by it, learned to trim the sails, and became knowledgeable about tides and weather conditions. She took her turn at the wheel and at the pump, and her father eventually made her the mate. Annie later married William Kingdon Slade, also of Appledore, when he was mate of the *Hawk*. William was a hard worker, ambitious, and a fine sailor, but he had never been to school, so Annie decided to educate him. At least in the early stages of their correspondence, she always returned his letters to him with corrections added. He never mastered the business letter, however, so Annie became their business manager, went to sea with William, raised their children on board, and continued to take her turn at the wheel and often served as mate or navigator or third hand. William progressed from being master of the ketch *Francis Beddoe* to the *Minnie Flossie*, a larger vessel, to owning a quarter share in the schooner *Alpha*, of which he was also master, to the schooner *Ulelia,* which was jointly owned with his brother-in-law, George Quance, who became master while William held the position of managing owner. Annie still kept the books and handled the correspondence, particularly to obtain cargoes, until she passed this responsibility over to her eldest son. In the meantime a house was purchased with the savings that Annie continued to accumulate, and the family had by now moved up to a comfortable middle-class position, due in large part to Annie's initiative and hard work. The relationship between Annie and William Slade was an ideal partnership, and William was unstinting in his praise of Annie for the unusual contribution she had made to the partnership. While Annie's story may be quite unusual, her son W. J. Slade concluded his tribute to his mother with the words: "There were many fine women who were ideal sailor's wives in Appledore" (Greenhill and Giffard).

It was not uncommon for women to go to sea as crew members in order to help the family business survive. Often there was no alternative until children were old enough to replace a working mother, or when there were no boys in the family. Such stories have largely gone unrecorded because neither the woman nor the family thought they were doing anything remarkable. The multiplicity of roles which women of many countries have played in connection with the sea are now fortunately recorded in a book edited by Jane Nadel-Klein and Dona Lee Davis and entitled *To Work and To Weep: Women in Fishing Economies.*

Many women who went to sea as captains' wives on whaling ships and trading ships seemed to have a natural fascination with navigation. It had the dual advantages of relieving boredom and giving the woman a more immediate involvement in the world of the ship. Captains often took advantage of their wives' natural curiosity about navigation by teaching them the intricacies of this science. Honor Matthews Earle, a former school teacher, became the official navigator on the whaler *Charles W. Morgan* between 1890 and 1906, when she sailed with her husband, Captain James A. M. Earle. She may already have studied navigation in school, but in any event she learned the practical aspects of it very quickly. When Captain William Mayhew and others of his crew became ill with smallpox aboard the whaler *Powhaten* in 1846, his wife Caroline became navigator and captain with the full support of the crew, who showered her with gifts, after the event, for a job well done. Lucy P. Vincent Smith, wife of Captain George A. Smith, in addition to using her sewing machine to make and repair sails for the ship, also learned navigation from her husband, and within a few lessons her calculations were identical to those of her husband. Thomazene (Tamzin) Williams, from an upper-middle-class shipping family of Appledore and wife of Captain Edward Williams, in a journal written in 1850 on her husband's ship from Newport to Quebec, shows an unusual knowledge of navigational methods and technical terms, undoubtedly learned in the cabin as she worked with her husband on the voyage. In 1897, Lloyd's of London awarded its prestigious Silver Medal to Mrs. Reed, wife of the captain of the *T. F. Oakes*, in recognition of the fact that she took command when her husband was ill, the first mate had died and several of the crew were dying of scurvy, and saved the ship. Captain Thomas Crapo pays high tribute to his wife, who was alone in the cabin when the brig *Kaluna* was smashed by a heavy sea. She quickly grabbed a lamp and, while standing waist-deep in water, held the lamp high so that the helmsman had a good view of the compass and could see to steer the ship. Her husband credited Caroline's presence of mind in saving the ship.

There are at least two outstanding instances in which women took charge of fully rigged clipper ships and brought them safely to port under the most trying circumstances. One story concerns Hannah Rebecca Crowell Burgess of Cape Cod, Massachusetts. In 1852 Hannah married William H. Burgess and went to sea with him in the clipper *Whirlwind*, taking lessons from her husband and studying books on navigation. William was so impressed with Hannah's competence that within one month he appointed her the official navigator. By 1856 William was given command of a larger ship, the *Challenger*. The ship was loaded at London for a trip to Callao, Peru. During the voyage William became seriously ill and was put to bed. Hannah consulted with Chief Mate Henry Winsor and agreed to take command of the ship and navigate it to the nearest large port, Valparaiso, Chile. Hannah had earlier recorded in her diary: "I like the *Challenger* very much. She is a much better sailor than the *Whirlwind* and 400 tons larger." All her skills were needed in the venture now unfolding as, with Henry Winsor's assistance, she daily recorded and charted the ship's route to its destination and attended to the needs of her dying husband. William died on the nineteenth day of his illness and Hannah refused to bury the body at sea. After twenty-three days in charge of the *Challenger*, Hannah arrived in Valparaiso on December 15, 1856. William's body was buried there and Hannah took passage home in the steamer *Bogota*, arranging the next year to have the body brought home for interment in West Sandwich Cemetery. Hannah was 21 years of age when she assumed command of the *Challenger*, and William was 27 at the time of his death. It is said that Hannah had fifty-six offers of marriage in subsequent years, but she had taken a vow never to marry again and died in 1917 at the age of 82.

An even more amazing story involved Mary Patten. She made two voyages with her husband, Captain Joshua A. Patten, on the clipper ship *Neptune's Car* between 1853 and 1856. On the first voyage she proved to be "uncommon handy about the ship," as Joshua noted in his log. She became so adept at navigation that it was felt she could easily qualify for a master's certificate. The second voyage was a race from New York to San Francisco around Cape Horn with two other clippers. Joshua was soon forced to place his first mate in irons because of his insubordination and abusive and careless behaviour; as a result of being on deck continuously for several days and nights, battling gale conditions, Joshua became seriously ill and was confined to bed. With the support of the second mate, Mary rallied the crew in a speech that sought their full support and assured them that she could get them safely to San Francisco. They were now approaching treacherous Cape Horn in storm conditions. Dressed in oilskins and wielding a speaker's trumpet, she commanded the crew and fought the elements for fifty days, during which time she slept in her clothes. As they hoisted one sail after another, each was stripped away by gale force winds for the first several days. Gradually the wind eased, allowing sails to be raised little by little.

As the captain recovered but before he got out of bed, he curiously released the first mate who immediately attacked Mary Patten. Hearing cries from the vicinity of the cabin, crew members rushed to Mary's assistance only to find the first mate prone on the floor and sporting a lump on his head. Having conquered Cape Horn, Mary Patten was not about to let a mere man overwhelm her! The first mate was again placed in irons. Sails were now crowded on for the race to San Francisco, but then the winds failed for ten full days before they reached their destination. The voyage took 136 days and was good for second place in the race. They arrived in San Francisco on November 15, 1856 (incidentally, just one month before Hannah Burgess!). The insurance company awarded Mary Patten $1,000 and showered her with accolades.

The amazing thing about this feat is that when Mary Patten took over command of the ship she was 19 years of age and four months pregnant! Her baby was born on March 10, 1857, and her husband, who never fully recovered from his illness, died three months afterward. Perhaps the ordeal had been too much for Mary Patten also; a combination of typhoid fever and tuberculosis ended her life at age 23.

These are just a few of the stories having to do with women and the sea. They can be multiplied over and over again by scores upon scores. Altogether they form a fitting context for a closer look at women and their ships. Many of these women were sponsored or mentored by fathers who saw their true worth or by husbands and brothers who regarded them as equals. There was in all of them a fierce inner strength and an independence of spirit that enabled them to make their mark and achieve a measure of true greatness.

CHAPTER TWO

Women and Ships

How this Project Originated

While doing research in the late 1980s for my book *For Love of a Woman: The Evans Family and a Perspective on Shipbuilding in Newfoundland*, I discovered the names of four women who were listed as shipowners in the microfiche edition of the *Maritime History Shipping Records.* It was, I thought, a curiosity but obviously an error. I made note of the four and determined to check them out at a later date to find out how such an error could creep into the records. Further checking turned up 97 names in the Newfoundland records, but I did not yet have access to the original records and that would require travel and additional work.

I was, at the time, a librarian at McGill University, and when I was awarded a sabbatical leave in 1993 I determined to make women shipowners a chapter in an intended book to be entitled *Sails of Newfoundland*, after the model of John P. Parker's book *Sails of the Maritimes.*

Research at Memorial University of Newfoundland in 1993 and 1994 yielded so much information on women shipowners that I determined it required a book of its own. Thus this effort.

In the very early stages of this research, a staff member at the Maritime History Archives in St. John's remarked to one of the Maritime historians that I was doing research on women shipowners, and he replied, "Oh, and has he found all two or three of them?" That response seemed reasonable at the time. It was common knowledge that Marie Smart Penny of Ramea had owned ships in the 1950s; and perhaps it was known that Ada Annie Petite of Mose Ambrose had owned ships in the 1930s and the 1940s and that Ada Annie's granddaughter, Debbie Petite, was a current shipowner at English Harbour West, but that would have been the extent of common knowledge up to that time. Women's involvement with ships was considered to be inconsequential. Some consciousness-raising was necessary.

I later had an opportunity to speak of my research to a group of Maritime historians and other interested parties at both Memorial University and McGill University, and the consensus was that sufficient material existed for a worthwhile research project.

Part of my sabbatical leave research consisted of visits to several Newfoundland communities where there had been a concentration of women shipowners. I thought it would be possible to gather some anecdotal information from the children or grandchildren of some of these women. I despaired of finding any hard information through diaries or journals or written family records.

Accordingly, in the spring of 1994 I visited Channel-Port aux Basques, Rose Blanche, Harbour Le Cou, Diamond Cove, Isle aux Morts, Harbour Breton, Mose Ambrose, English Harbour West, Belleoram, Burin, Grand Bank, Fortune, Glovertown, Carbonear, Spaniard's Bay, Western Bay and Harbour Grace. I started my visits in Rose Blanche, dropping in at the post office and showing the clerk a list of some of the women's names. I was literally pointed to a house where one of these women was still living! Unfortunately she was suffering from memory loss, but the visit with her daughter and son-in-law was worthwhile. At Rose Blanche, Harbour Le Cou and Harbour Breton, I found six of these women shipowners still living and two of the husbands whose wives had been shipowners.

Ownership by these six women primarily covered the period 1943 to 1962, but the period of ownership by Maud Billard of Harbour

Le Cou covered most of the years between 1947 and 1983. Maud owned two ships, the latter one a longliner. The five other women were Myrtle Hatcher, Mary Elizabeth Hannam and Clara Theresa Hardy of Rose Blanche, Mary Yarn of Mose Ambrose, and Lillian Martha Hynes of Harbour Breton. These women owned one ship each. Violet Blanche Butt, wife of Simeon, of Harbour Le Cou had owned a ship through the 1930s and 1940s. I visited Simeon in 1994. The other husband was Henry Hatcher.

Most of these women were quite elderly when I visited them in 1994. Some were suffering from memory loss and unable to give detailed information. But three visits yielded good solid data nevertheless.

Myrtle and Henry Hatcher of Rose Blanche were 85 and 93 when I visited them on May 26, 1994. They were mentally alert, charming and very hospitable. Myrtle Riggs had been born in Anderson's Cove, Fortune Bay. She began knitting for NONIA (the Newfoundland Outport Nursing & Industrial Association) when she was 14 years old, that is, in 1923. NONIA had been formed by Lady Constance Harris, whose husband was the governor, and its aim was to place nurse/midwives in Newfoundland outports in order to improve health care. The program was later expanded by Lady Elsie Allardyce to include knitting and weaving, the profits from which supported these community nurses. Dr. Wilfred Grenfell sent a trained weaver to Poole's Cove, and Myrtle Riggs went there for special instruction. For 15 years she knit, did weaving and hooked mats for NONIA and was designated one of six persons to do fancy knitting and crocheting. When she had difficulty with the intricate and demanding requirements of the instructor, she would dream during the following night about how to do it and that always resolved the difficulty! In 1942, at age 33, she came to Rose Blanche as housekeeper for Henry Hatcher, a successful fisherman. Henry's wife Elizabeth had died and left four children. Myrtle looked after the children, "did a little flake work at the beginning" (i.e., curing fish), and cooked and washed for the sharemen. She said, "They were good times. No one was rich; everybody was friendly and helped one another." Myrtle eventually married Henry and they had three children. Myrtle and Henry bought the 23-ton *Minnie & Joan* in 1946 from Robert Allan Newman, merchant at Petites. Myrtle and Henry were joint owners; each owned 32 shares, and with a successful fishery they paid for the vessel in 10 months and operated it for 10 years. When I asked how Myrtle became a joint owner of their ship, Henry explained that it was "for the protection of my wife." Myrtle was still living in 2005.

Mary Boone was born in Bareneed on November 5, 1912. Before moving to Mose Ambrose in 1949, she taught school for more than 20 years at Daniel's Harbour, Bellburns, Brig Bay, Savage Cove, Herring Neck, Shoal Cove West, and then moved to St. John's to teach at Springdale Street School and the Anglican orphanage. Chesley Yarn was a merchant in Mose Ambrose. His wife Stella was from Harbour Grace. She had a tubercular knee at age 17 and wore an artificial leg. By the time Mary Boone moved to Mose Ambrose, Chesley had long been a widower. They married in 1954. Chesley was a merchant with a business in Mose Ambrose, and with Mary's cooperation they started a new business in nearby English Harbour West for which Mary became responsible. She drove her car every day to the new location. Chesley had gone into "the ship business" about 1952 and operated the *Hilda M. Tibbo* as a passenger boat for several years until it burned. For six years Mary owned the 15-ton *Wagaymack* in her fishing business. When Chesley bought the vessel in 1956 the ship registry noted: "Mary Yarn of Mose Ambrose, NF., the person to whom the management of the vessel is entrusted by or on behalf of the owner. Advice under the hand of Chesley Yarn received 17 September 1956." In other words, Mary was appointed as managing owner of the vessel. In Mary's own words: "He made me part owner to provide for me as a partner." She operated the business in English Harbour West for 12 years and then assisted with Chesley's business at Mose Ambrose. When I mentioned to Mary that my reason for visiting her was that she had been a shipowner, she said she had not owned a ship. Then I reminded her of the *Wagaymack* and she said, "Oh, that was a boat. I thought you meant a big ship." She was not about to take undue credit. The fact remains that it was a decked vessel and therefore required to be registered at St. John's. In praise of her husband, who had only a Grade 9 education, Mary said: "I had the book learning, but Chesley was able to go around me in everything else. He was keen. An adding machine was no good to him; it just slowed him down." Mary was an intelligent, well-read, well-spoken woman, a gracious host for my visit with her in 1994. The bookshelves in her living room held an array of good books, many of them recent publications, as well as several current magazines.

Violet Blanche Hannam was born on January 25, 1890, and married James Horwood, a merchant. When her husband died, Violet was left with five children and everything was tied up in the Horwood estate, from which she received nothing. In desperate straits, she came with her children to live with Simeon Butt of Harbour Le Cou as a housekeeper. In those days Simeon was fishing in "skiffs, two-masters with no boom, a little bigger than a snapper." They married on December 8, 1931, though Violet was 14 years his senior. Violet did all the work in the home and she spread the fish and ensured that it was dried (cured) properly. She had a better education than Simeon so she looked after all the money and "all the trade." But they consulted on all decisions. "One didn't do anything without clearing it with the other, without asking." On January 14, 1935, Simeon bought the 15-ton *Agnes R* from the merchant Robert Newman. This was an old vessel. It was built by Frank Strickland of North Arm, La Poile, in 1916 and rebuilt in 1934. As testimony to the successful fishery of that time, in less than three months Simeon had the vessel paid for, and on April 1, 1935, he had it registered in the name of Violet Blanche Butt, his wife. At the same time he had his house put in his wife's name. When I asked him why he did this, he replied: "I did it in recognition of her contribution, and I never wanted her to know poverty again." She died in 1977; Simeon was still living in 1994.

On this same visit I met and briefly interviewed Debbie Petite, Manager of J. Petite & Sons Ltd. at English Harbour West. Debbie is the daughter of Gordon Petite and the granddaughter of Ada Annie Petite, who operated the family business in Mose Ambrose for several years after her husband Harry was lost at sea. I have corresponded with Debbie Petite since that visit. As manager of the business, she is now a shipowner also.

These brief stories give some indication that women were involved with ships and the fishing business in a substantial way. But they represent only the tip of the iceberg. Women were involved in the ship business in even more substantial ways, as we shall see later.

In the almost 20 places I visited in 1994, I was in search of information on 114 women who were involved in the ship business from 1823 to 1983: 7 women in Channel-Port aux Basques, 17 in Rose Blanche, 6 in Harbour Le Cou, 6 in Harbour Breton, 4 in Mose Ambrose, 1 in English Harbour West, 6 in Belleoram, 2 in Glovertown, 12 in Burin, 21 in Grand Bank, 5 in Fortune, 17 in Carbonear, 2 in Spaniard's Bay, and 8 in Harbour Grace. These were the areas with the largest concentrations of women shipowners and it was convenient for the travel time available. I feel fortunate to have been able to visit many of the last people to be involved in the merchant/fishermen salt-cod business.

Women in about 130 places in Newfoundland were designated shipowners in the ship registers. I sent letters and questionnaires to people in several of these communities in an effort to gather information about these women and combed through the records in local libraries for elusive biographical data. It was impossible to gather all the material one would desire.

Women as Shipowners

So few hard data are available on women for the period covered in this study, roughly 1675 to 1975, that we can only guess at the details in many instances. In this 300-year period I have been able to identify approximately 1200 to 1500 women who were involved substantially in the ship business in the four Atlantic provinces and Quebec. In Newfoundland I have gone back to the earliest census (1675), and I have consulted court records and a variety of other sources in order to get a clearer idea of the durable woman who emerges from that misty past.

To give the reader a better idea of the breadth of women's involvement in the ship business, below is a list of the various categories I have been able to identify from the records of all five provinces, of England, and of the United States:

- Women as sole owners of ships
- Women as joint owners of ships
- Women designated as co-partners in trade
- Women designated as managing owners of ships
- Women appointing men (husbands) as managing owner
- Men appointing women as managing owner
- Women who bought shares in ships
- Women who inherited shares in ships
- Minors who were designated to inherit shares in ships (often with mothers)
- Women as mortgagees for ships
- Women who took mortgages
- Women owners who did not have a mortgage but men to whom they sold the ship had to take a mortgage
- Women who inherited ships and sold them within one year or so
- Women who inherited ships and apparently remained active in the business
- Widows who remained in the business after the husband's death
- Widows who bought ships
- Women appointed as sole executrix of a will
- Women appointed as executresses (i.e. women only)
- Women appointed as administratrixes
- Women appointed as one executrix among two or more persons
- Women designated as executors
- Women buying insurance on ships
- A woman empowering the captain to sell her ship

Women dying with a will

Women dying intestate

Women naming ships after themselves

Women's occupations as designated in the ship registers

Women's ships involved in the sealing business

Women involved in the whaling business

Married women owners before 1876 (i.e. before Women's Property Acts)

Women selling ships directly to women

Wives selling to husbands and vice versa

Women owning in partnership with women

Couples selling directly to couples

Women owning with husbands

Women owning with husbands and others

Women owning with men (brothers, fathers, general – not husbands)

Women apparently "staking" planters, traders and mariners to build ships

Women as navigators

Women as captains and mates

Women as shipbuilders

Hints of scandal involving women's ships

Women property owners, by purchase and inheritance, including married women

Some will ask the question: Why did women own ships? But the counter-question, based on the evidence, is equally relevant: Why not? There was a natural transition from owning property to owning a family fishing business to owning small boats for an inshore fishery to owning ships for fishing, trading and sea-going ventures and to buying shares in ships for profit-making. Perhaps it all began with "the first woman" inheriting a ship through her husband who drowned at sea, and realizing that the family business must continue for the survival of her own children, and knowing that she has the schooling to keep the books and market the fish and the knowledge and experience to cure it for the greatest profit. She keeps her ship and operates her business successfully and proves to other women that it can be done.

Perhaps there were almost as many reasons for women owning ships as there were women involved in the business. Fathers, seeing their daughters' potential for life on the sea, encouraged them and even on occasion put their daughters' careers ahead of their sons'. Husbands and families needed the expertise and the labour of wife and mother, so she became involved in the family enterprise in order to ensure survival. There is evidence, too, that some women were role models for other women, and this is particularly apparent in some south coast communities of Newfoundland, where there were concentrations of women shipowners, and at Saint John, New Brunswick, where women modeled share-purchasing for succeeding generations. The women of New Brunswick were daring speculators. Their money was almost always invested in new large ships, and many of these were built to be sailed across the ocean to Europe and sold for a quick return on investment. This same pattern was also true of Prince Edward Island women but on a much more limited scale.

We must not underestimate the initiative of women themselves in buying ships and shares in ships. The ship, as the workhorse of its day, as a major means of transportation, as a great carrier to distant and exotic ports, represented adventure, romance, security and profit for those who invested in them. Women were drawn to the sea as much as men were. Many women must have liked everything that had to do with the sea as much as men did. Many of them were members of fisher families, of merchant families, of lumber-producing families, and the ship was vital to their way of life. Many women had money, and it was a natural act to invest their money for profit.

There was also another reason why women owned ships: to protect the family's assets. The husband may have gone into debt to a merchant and as a result registered the ship in his wife's name until the economic situation changed. In Quebec, many women had a civil contract in addition to the religious marriage, and by this contract their goods and assets were legally separate from their husbands' goods and assets, so that if the husband went bankrupt or fell on hard times the family could live on the wife's assets until the husband recovered financially. A businessman may have divided the family's assets by registering some of the ships in his wife's name without any particular involvement in the firm's affairs by his wife. The latter appears to have been a practice of some merchants and may have been a hedge against possible bankruptcy. It is difficult to identify such cases from the skimpy information given in the ship registers, but I would wager that these instances are the exception rather than the rule.

Was women's ownership real? The evidence suggests a resounding "Yes." Apart from the few instances cited previously, the records show that fewer than 100 women in the five Atlantic provinces inherited a ship and sold it within one or two years, opting not to carry on the venture. The vast majority of women held on to inherited ships and pursued the business, several for many years. Quite apart from that, women were active in buying ships or even in having ships built for their ventures. I have a copy of the bill of sale dated May 30, 1872, in which Cecilia Brown, widow of Stromness in the County of Orkney, executress to the late William Carpenter Brown, bought 16 shares in the 131-ton brigantine *Annie,* which had been built in Murray Harbour in Prince Edward Island in 1867. Cecilia paid 331 pounds 15 shillings sterling for the shares and held them for 14 years, her agent selling them on behalf of her estate when she died in February 1886. So women's ownership was real and is thoroughly documented.

Other appropriate questions are: How did women buy ships? Where did they get the money? Many women had their own money, especially those who came from well-to-do merchant or planter families, and many of these brought this money into their marriages. The wills of Newfoundland women and the Crown Lands Registry, both of which I have searched, reveal that many women were rich in lands, houses, property and bank accounts. Grace Anderson's will of 1856 leaves 230 pounds sterling to her niece. The will of Catherine Mahar, widow of St. John's, in 1884 refers to three dwelling houses, a shop on Water Street, and property on Forest Road, and makes provision for her grandchildren's "maintenance and education" until they marry or become of age. The will of Ellen Dawe in 1885 refers to six houses, which she leaves to various family members, and additional property and annual financial provision for her grandchildren. Mary Ann Voisy, widow, in 1912 left four houses, two of these with two tenements each. And Martha Nichols' will of 1918 states that "my money in the Bank is intended by my wish when I part this world for my husband and children, my husband to draw right away signed by myself… Also money left me in any other way my wish is for my husband and children to have it."

Several women in Newfoundland were sawmill operators and held timber and mining licences, and even licences for a salmon fishery. Mary Winsor, widow of James J. Winsor, merchant of St. John's and Exploits, Burnt Islands, after her husband's death in 1889 became a co-partner with David Lewis in the very large sawmill at Dominion Point, Peterview, and continued in this capacity until the mill closed in 1892. That mill had been in existence since 1869 and was the catalyst for the settlement of the entire Botwood area. Thomas Raymond Murphy started a sawmill at Glenwood in 1927, and when he died in 1942 his wife Katherine partnered with Zipporah Steele, the local telegraph operator; they operated the mill until 1966 when they sold it. Prior to her marriage, Katherine had been a bookkeeper at the Crosbie Hotel in St. John's for 17 years.

Mrs. Martha Cottman, "a subject of Great Britain," in 1751 applied for a grant of the North-east Arm of Placentia to carry on a salmon fishery "and the more effectively to complete the same." The record continues: "…that she desires the property of the grounds from the West Side of the two Islands commonly called Salmon House Island and Crasous Island and for half a mile around the same… and the peaceably and quiet possession of the same without the least hindrance or molestation in the enjoyment thereof." She was covering all the bases! E. Brown, J.P., awarded Martha Cottman the grant "by virtue of the power and authority to me given by

Admiral...Drake." The widow of Joseph Hornett (probably Ann) held a licence for a salmon fishery on the Campbellton River from at least 1861 to 1875, inheriting it from her husband and passing it on to her son. It is probable that Charlotte Rhodes, widow of Richard Rhodes of St. John's, who was awarded 29 acres, 3 roods, and 8 perchas of land at Salmon Brook, Northwest Arm, Clode Sound, on January 3, 1898, at a cost of $8.95, was also intending to pursue a salmon fishery.

Women were also awarded mining licences. Mary J. Verron, widow of Placentia, sometime before 1883, was awarded a licence "to Occupy and also to dig, mine and work, all and every the ores, mines and minerals contained in the Track, Piece and Parcel of land..." On October 12, 1883, Mary transferred by deed one half of her interest in the licence to Daniel J. Henderson, Gentleman, of St. John's. The record continues: "We grant them all the minerals with the exception of gold...one square mile...South-east Arm of Placentia." In 1895 they jointly held 45 acres, including Verron's Island, so obviously Mary had the distinction of naming the island after herself. In 1901 Elizabeth Browning of Rogue's Harbour was granted mining licences at South-West Arm (Point Leamington) and Seal Bay, Notre Dame Bay. Selina S. Sullivan of Presque, Placentia Bay, "acquired certain mining rights from the estate of C. F. Bennett on the West side of Placentia Bay...she has now applied for a Fee Simple mining grant...at Red Cove and Great Bonia...(and) on Presque Arm, St. Kyran's." This was in 1904. Annie Oxley in 1920 held seven mining licences in the Nichollsville, District of St. George, area. She was now applying for a grant in fee simple (given ownership of the land with unrestricted rights of disposition), and it was noted that she had spent more than $6,000 on each site and had "fully complied (with the terms of) the grant." It is noted further that she was a native of Washington in the County of Durham, England. It appears that she may have been the widow of John J. Oxley. Annie Oxley was also a joint holder of a timber licence, as described later. In 1923 two women, Elizabeth M. Tobin and Mary A. Tobin of St. John's, "did obtain a 99-year lease for two mining locations between Frost's

Cove, Moreton's Harbour, and Western Head, North-west Island" and pledged to spend $6,000 on the project.

And then there were timber licences. On June 22, 1908, Jane Taylor of St. John's was awarded 136 square miles in several areas around Sandy Lake and near Mount Seemore, Seemore Lake and Birchy Ponds. She sold the licence just six months later, on December 9, 1908, to Frank A. Begent. William H. Taylor of St. John's was also buying timber licences at the time and may have been a relative of Jane. One of six names involved with a timber licence at Crabb's River Island from Bay St. George when the licence was sold in 1906 was that of Annie Oxley. The names of John J. Oxley and William H. Taylor appeared on the sale document. Annie Hender was a partner with Samuel J. Foote in a timber licence at St. George's River in the District of St. George's on June 18, 1918. The area consisted of 40 square miles, bounded on three sides by the Reid Newfoundland Co. and on the fourth side by the St. George's River.

The Crown Lands Registers contain many references to women purchasing property, lands and houses. These were widows, spinsters and married women. One was a shopkeeper, another a seamstress. Two or more women purchased land together. Women purchased with husbands and with men other than their husbands. I was constantly surprised by the number of women noted in these transactions.

It is not surprising then that when James Evans, merchant of Adam's Cove, died in 1919, his wife formally announced that she would carry on the business. The following notice appeared in the *Evening Telegram* on October 18, 1919: "Notice is hereby given that the business carried on during his lifetime by James Evans, late of Adam's Cove, merchant, deceased, will be continued by his widow, Mary Ann Evans, who will be responsible for all the debts due or to become due in respect of said business and to whom all persons indebted to the said business are required to make immediate payment. Dated October 17, 1919. Eastern Trust Co. Administrators Cst.a. of the Estate of James Evans, deceased." This was a substantial business in Conception Bay dealing with large cargoes of lumber and coal, as well as a wide range of general merchandise, and there are many references to it in the newspapers of that period.

And so, back to ships. Some women bought or were given as little as one share in a new ship that was being built, and this share was used to produce a profit for the individual involved. That invested money usually grew, though sometimes there were losses. When I lived in Grand Bank in the mid-1960s I talked several times with Charles Forward, one of the principals involved in the firm of Forward and Tibbo. Mr. Forward was then an elderly man and he liked to talk about the fishing business on the Grand Banks and his successes in concert with his partner Felix Tibbo. Their wives, Mary Florence Forward and Eleanor Tibbo, were also shipowners. Mr. Forward showed me both his and his wife's personal financial record books. Mary Florence was then deceased. Their records were entirely separate and Mr. Forward described to me in detail how his wife had divided her financial holdings among family members, friends and charities in preparation for her impending death, and how he had already done the same thing in his separate will. I am sure that the same thing was true of Eleanor and Felix Tibbo and countless women in Grand Bank and in the entire Atlantic region.

We should not underestimate the initiative of women even in poorer families in making and saving money towards the survival of the family. Like many a woman in Newfoundland, my mother had a special container for "egg money," and all earnings from selling eggs, milk, cream, berries, and special baking went into this container. My mother-in-law loaned such money to young people going away to find work and was always repaid. These small earnings in families often made the difference between survival and poverty. A recently published book, *Egg Money Quilts* by Eleanor Burns, pays tribute to three of the author's female forebears and their industry and initiative in using "egg money" to meet family needs.

Women with money bought ships and shares in ships, but they also became mortgagees, securing loans for others to buy ships and charging interest to make a profit. The fact, for example, that women were sometimes designated "co-partners in trade" emphasizes the reality that women had money, made money and shared in the profits from fishing and trading ventures.

Women and Property

It would be appropriate at this point to consider the law and practices that governed women's inheritance of property, particularly during the eighteenth and early nineteenth centuries. There appears to have been some erosion of rights in practice from the seventeenth and early eighteenth centuries, which is ironic in view of the fact that it was a British queen, not a king, who ruled through much of the period under study.

In his thesis, *Economic and Social Relations of Production on the North-East Coast of Newfoundland, with special reference to Conception Bay, 1785-1855*, Sean Cadigan quotes some examples which seem to reflect a pessimistic view of how women actually fared in cases where they inherited property from their husbands. He writes: "Widows usually inherited little property from their husbands' estates. Those who did, like Mary Sheppard in 1788, were not allowed to alienate what little women did inherit from the husband's patriarchal line." It is true that many husbands' wills tended to disinherit the wife from the estate in the event that she remarried, but perhaps because of the preeminence of the family as an economic unit and the wish to guarantee that it would remain so in law; also, if the wife remarried, she joined another economic unit which might in some cases have been in competition with her children's. I would interpret differently some of the examples Cadigan gives because many widows seem to have been adversely affected by contentious, quarrelsome sons and stepsons or sons born of other marriages. Even in Mary Sheppard's case, she was obligated only to pass on her late husband's "goods" to her son Adam after she died, but Adam died before she did, so "the watch and feather bed," which Adam had apparently taken, were to be returned to Mary and "the said Adam's Heirs to pay the sum of 6 pounds to said Petitioner for the use of what goods he enjoyed during his Life and which belonged to his Mother..."

The example of Ann Mugford, wife of the deceased Robert Mugford, planter at Port de Grave, was a case in which Ann violated the terms of her husband's will, which stated that she had inherited his property for her natural life and then it was to pass to their eldest daughter Frances. Before her death in 1813, Ann Mugford had sold the property to John Walsh, and in 1817, Nicholas Newell, a planter in St. John's, petitioned the court to recover the property since he was married to Frances. The Chief Justice acceded to this request.

There is no doubt that the rules governing inherited property were patriarchal, but in actual fact, when good relationships existed between the widow and her children (as was usual), the family as an economic unit survived and flourished and the widow often continued to play a leadership role in guiding the family fortunes. In fact, Cadigan pays tribute to women "who could take on important economic roles as estate managers and temporary household heads throughout the Anglo-American world in this period," and he acknowledges that women like Jane Cook of Harbour Grace acted for other women in leasing property to planters and handling women's accounts with merchants and that "women managed property and accounts left to them by their husbands."

Cadigan builds a very strong case for the way in which "...Newfoundland households integrated women's work into commodity production..." as contrasted with Upper Canada. From the very earliest times up to the 1950s, as long as the salt-fish trade existed (i.e., the preparation of salted, dried codfish for foreign markets), women and men worked together as co-producers in family units. Every other activity in the fisherman's household was subordinate to the salt-fish activity. Women had to be prepared to set aside everything else they were doing when the fish arrived, for they were acknowledged to be the experts in salting and curing the fish to meet the strict demands of the market.

Nowhere is this specific activity better described than in Hilda Chaulk Murray's *More than Fifty Percent: Woman's Life in a Newfoundland Outport, 1900-1950*. She writes: "In Elliston, in the period prior to 1950, the women were full participants with their menfolk in wresting a living from the sea and land, and were directly involved with all the economic, as well as social activities, in the community. The role of the fisherman's wife was completely intertwined with that of her husband." Women participated fully in every aspect of the fishing operation except in the catching of fish. Even the making of sails for the boats was a cooperative venture: the men cut the sails, the women sewed them, and the men roped them. When accounts were settled up in the fall, the money was brought home to the woman who "handled the cash and managed the day to day running of the household."

The routine that Murray described for Elliston was not entirely uniform throughout Newfoundland. For example, when the fish for the large business firms was brought into communities like Grand Bank it was transported by horse and cart to one of the beaches and placed in large piles where the curing process became the exclusive responsibility of the "beach women." They worked in teams with a boss woman in charge of nine others. Their distinctive attire was characterized by white aprons and white sunbonnets. The fish was spread each day for a month or more until it was dry, then graded, then moved into sheds for inspection by the culler. It must have been one of the striking early examples of assembly-line production.

The St. John's newspaper *Free Press* of November 25, 1913, gives a wonderful example of the cooperation between husband and wife in a fishing venture: "Patrick Tobin of Gray Islands holds the record as an individual fisherman in Newfoundland for having made the biggest bill in the shortest season. A mercantile firm paid him $1,050 for his season's work – July 7 to September 15 – he landed 160 quintals of cod, fishing 'cross-handed.' The fish was split and cured by Tobin's wife and self, who also made the oil ready for market. This is the largest individual bill to be made by any fisherman in Newfoundland." Though Tobin fished "cross-handed," i.e., alone, his wife on shore split and cured all the fish and made the oil ready for market. These were important skills which ensured the family's survival. There is no mention of children helping with the work, but the example serves to confirm the conclusions of other similar sources.

There are, as well, much earlier examples of women's involvement in the fishery in a hands-on manner. Noel H. Bowen, writing in 1855, decries the lack of women along the Labrador coast, citing an 1852 report that there were 364 men settled on the coast and only 62 women, suggesting that the situation should be remedied by emigration or some other plan, and arguing that "the more the coast is settled the better are its natural resources developed." He writes further: "The fisherman's wants are few, and he can easily support a wife; moreover she could assist him in his calling as much as one of his own sex, and it often happens that the larger a fisherman's family becomes, the better are his prospects..." But Ephraim Tucker, an American who had spent five months on a voyage to the Labrador in 1838, actually saw women at work in the fishery: "But they engage in the hard and laborious toils of fishing with as much zeal and activity as the males. When the salmon and trout fishing commences, the women and children employ themselves assiduously in the sport, and are often out night and day while the season of this fishery lasts. At the fish stands, while the cod fishery is in the full tide of operation, the women are seen among the most constant and dextrous in dressing the fish, thrown up by the fishermen. Some of these females will dress two or three thousand fish in a single day..." While it may have been "sport" to Tucker, who was at the Labrador for his health, it was mighty hard work for women, men and children. What is most interesting about his words, however, is the confirmation that women were the experts at handling, curing and "dressing" the fish.

The merchants of Newfoundland who used (and preferred) the barter or credit or "truck" system found it in their interests to maintain the patriarchal nature of the fishing business. They preferred to deal directly with the male head of the household. They were at least consistent in this, for in cases where they sued the family, it was always one man or a small group of two or three men, but never a woman. Since the headquarters for the settling of accounts was not always in the same community where the fisherman lived, the merchant seldom had to deal with aggrieved women. In at least one case in which a merchant had to deal with a woman, she was victorious. Mary Reed in 1835 worked for a planter, Thomas Davis, who claimed not to have sufficient money to pay her. When Reed charged that Charles Nuttall, a merchant at Harbour Grace, had received Davis' fish and oil from the voyage, Nuttall was ordered by the court to pay Reed both her wages and damages.

Dona Lee Davis, who conducted a survey in 1978 of women in a south-west coast Newfoundland fishing village and published the results in the book entitled *Blood and Nerves: An Ethnographic Focus on Menopause*, writes: "Women feel particularly free to cause a spectacle or behave rudely when their family or household has been threatened or wronged." She found in her research that men will often defer to women's desire and ability to put things right. In the words of one of her women respondents: "A man has to be careful but a woman can say exactly what she likes."

In any event, the family's struggle for survival in what was often a harsh environment, and always against formidable odds, required a strong partnership between men and women in households. Male authority was routinely challenged and compromised for a higher purpose: the family's very survival. The result was a family solidarity that has persisted beyond what might have been expected. This is not to imply that marriages were always ideal relationships, but at least in the earlier periods personal self-fulfilment for both parties, and especially for the woman, was harshly subordinate to existing and surviving.

The written law and unwritten rules and practices have conspired through the centuries to treat women differently from men. A woman has had to fight harder and sometimes "behave rudely" in order to ensure that her case would be considered. Women occupied a dependent status in law for a very long time and were not even defined in certain jurisdictions as "qualified persons" for certain jobs. Power and responsibility were somehow connected with property ownership, and most property was vested in the name of men. Lack of power and responsibility combined with unjustified prejudices about women as inferior and incompetent, as ornaments for men's pleasure and convenience, to create practices of widespread discrimination against women. Marriage became, in the thinking of many men, a means of licensed control over women because the woman's identity was absorbed into their own. This false view of marriage has been rightly viewed as a form of bondage.

The English Common Law gave a husband almost unlimited rights over property belonging to his wife, which represented, in fact, an erosion of certain rights and freedoms that women in Europe had held in the pre-industrial period; women gradually became more associated with household, domestic, social and family matters in later periods. And all during the reign of a British queen!

Newfoundland became an official colony of Great Britain in 1824, and by 1837 had officially adopted English criminal law, which remained in effect until 1949 when Newfoundland became a province of Canada. English criminal law actually had the effect of curtailing more liberal practices with respect to women's ownership of property and management of estates, which had prevailed in the colony prior to this period.

Advent of Married Women's Property Acts

While the rights of widows and single women were approximately the same as men, except for civic rights, the rights of married women were severely curtailed in law. The battles for reform were being fought in England, both by women and by men who supported them. Two of the most famous proponents of reform were Elizabeth Barrett Browning and John Stuart Mill. Few of the supporters of reform were happy with the limited success achieved in the Married Women's Property Act of 1870, and the battle continued until a more liberal act was passed in England in 1882, the "Magna Carta" for women. A few quotes from *Registration of British Ships* published in 1896 are appropriate here: "As regards married women, a reference to the Married Women's Property Act, 1882...will show under what conditions they are in this country capable of holding ships...the power of a married woman to transfer shares is not absolute. Since the Married Women's Property Acts have been passed, the titles of married women to shares in shipping have frequently been recorded... A registered mortgagee has an absolute power of sale subject to the conditions imposed by Section 35."

In 1834, Newfoundland had passed the Real Chattels Act, which declared all landed property to be "Chattels Real" and gave responsibility for settling affairs to an appointed executor or administrator, subject to the direction of the Supreme Court of Newfoundland. Historiographers are still trying to determine the full impact of this unique piece of legislation. However, it is worthy of note that following the passing of this Act, women were appointed as executors and administrators (though they are usually called executrixes or administratrixes) of estates, and that this procedure usually passed quickly through the Supreme Court. From the ship registers of all the Atlantic provinces, it is clear that the Supreme Court, which handled estate settlements after death, treated them in a straightforward manner and with reasonable speed. Probate was often granted within a month or two of the owner's death. In cases where the owner left a will with a designated executor (often the owner's wife and/or brother or business associate), the same time-frame is observed. In cases where there was no will, the Supreme Court granted "Letters of Administration," usually to the wife, and she either sold the ship quickly or continued with the family business, often becoming the business head until one of her sons was old enough to assume responsibility.

In 1876, Newfoundland passed the Married Women's Property Act, based on the British Act of 1870. By this Act, property which a married woman acquired was designated as her own and could be used independently of her husband. A married woman could now carry on a business, occupation or trade separately from her husband. She could register in a joint stock company; she could sue and be sued; and she was now responsible for debts which she accumulated. Not until this Act was amended in 1883 did married women acquire the legal status of a single woman with respect to owning property. This Act was further amended in 1895 to deal with contracts, court action and wills.

Similar legislation came a little later in the Maritime provinces. The Married Women's Property Act was passed in 1884 in Nova Scotia, 1895 in New Brunswick, and 1896 in Prince Edward Island. It was not until 1964 that a similar Act was passed in Quebec. The Civil Code, derived from the old French Law, had operated in Quebec since the sixteenth century, when the territory was known as New France. It is very different from the Common Law of England. The Civil Law is codified as opposed to the largely unwritten Common Law. Marital status, including a woman's right to own property, was strictly defined under the Quebec Code. Many of a woman's legal rights were ceded to her husband at the time of marriage, including property rights. While single women and widows held essentially the same legal rights as men in Quebec, married women were discriminated against simply because they were married. There was one saving grace possible for married women under the Civil Code; it was possible to have a legal marriage contract separate from the religious ceremony of marriage, and many women who held property or possessions at the time of marriage entered into such contracts, as is evident in the Quebec ship registers.

In spite of what the law in all the colonies said about married women and restrictions on property ownership, there is clear evidence that married women owned ships and shares in ships well before the Married Women's Property Acts were proclaimed into law. We will look at the details later, and a summary of these data will suffice for now. In Newfoundland married women owned shares in ships in 1844, 1853, 1854 and 1866, as these ships were registered. That first date, 1844, was 32 years before the Married Women's Property Act became law in Newfoundland! *The Conception Bay Plantation Book* of 1805, with antecedents dating well back into the 1640s, demonstrates that married women owned property, boats, fishing premises, and possibly ships, well before 1800.

Nova Scotian married women owned ships in 1877 and 1884; the first date was seven years before the Act was proclaimed into law. In New Brunswick married women owned shares in ships that were registered in 1860, 1870 and 1872; and in 1873 the floodgates opened and many married women began to purchase shares despite the fact that the Married Women's Property Act did not come into effect until 1895. In Prince Edward Island a married woman held shares in a ship as early as 1827, and there was not another until 1895. The relevant dates for Quebec married women were from 1827 to 1903; there were so many, in fact, that either they all held civil marriage contracts or they simply ignored the existing law. The number of women holding shares in ships constitutes clear evidence that married women were following a pattern established in the pre-industrial period and did not wait for the law to affirm their right to own property.

Women's Occupations in Canadian Ship Registers

It is almost certainly true that occupation was self-designated and that the registrar from earliest times recorded what he was told by the person registering the ship. The registrar was male and most of those who came to register their ships were male. Women were also conditioned by the norms of early society to designate themselves as widows, spinsters or married women. Were it not for a scrapbook kept by Quebec registrars from 1850 to 1958, and which I found by happenstance in Ottawa, we might never have had a public pronouncement on the issue. It is clearly stated in a letter from the Department of Marine and Fisheries, dated December 30, 1949: "Where a woman's name appears as owner or mortgagee, she should be described as Spinster, Married Woman, or Widow, as the case may be." This directive was, of course, a complete reversal of a practice which had evolved over many years, and I expect that the bureaucrat who penned these words was unaware of what registrars in the provinces had been doing. Did his words, on the other hand, represent an attempt to reverse the fortunes of women in that society, or were they purely personal?

There is a section in the Quebec scrapbook entitled "Description of ownership title of a woman" and a note to the effect that regarding mortgages, the "marriage contract may be used instead of Last Will & Testament in case of death." There is a further note on "Married Women, Legal restrictions as to their right to hold or transfer ship property not the business of the Registrar." There is a letter from the Deputy Minister of Marine in Ottawa dated September 15, 1880, indicating that if Madame Beaupre wishes to be the mortgagee for her husband's, Maxime Beaupre's ship, that is really of no concern to the registrar's office; if she wishes to risk her money that is her prerogative. A further section of the scrapbook is indexed as "Registered owner may sell to his wife," suggesting only that the wife would first have to file a Declaration of Ownership and a Declaration of Assets. These documents were to be sworn before a notary public. A further index entry for 1949 is entitled "Title of Married Woman" and refers to a section where a woman had been registered as a "Mariner" but this should now be changed to "Spinster"; another woman had been designated as a "Manager" but this should now be changed to "Ship's Husband." I wonder if the registrar saw the irony in the latter dictate. There is an added communication signed by Elizabeth Descheneaux's attorney indicating that she had originally been called a married woman and this has now been changed to spinster.

There is some evidence that existing records were "corrected" by a registrar who served after the time of the original registration. There is an instance where "Stenographer" was marked over and "Spinster" was written in by a later hand. There is a case where "Hotel Proprietor" was crossed out and "Married Woman" was written in.

And yet another case where "Housewife" was crossed out and "Married Woman" was written in. But only about half a dozen such instances were found in the ship registers of all five provinces.

So what are the facts related to designated occupation? The earliest women shipowners, dating from the early 1800s, were widows, but it was not long before single women made an appearance, and then married women. In Newfoundland the first widow to register a ship was Ann McCarthy of Crocker's Cove, Conception Bay. She registered the ship *Susan* at St. John's in 1822 as sole owner. The ship had been built at Moreton's Harbour by Richard Horwood. Modifications must have been made to the ship subsequently since it was re-registered in 1825 when it was noted that the ship was engaged in the Labrador fishery, and Joseph Taylor was the master or captain. Twenty years later, in 1842, Mary Foley Morris, a single woman of St. John's, registered three ships: the *John and Mary*, the *Nancy*, and the *Relief*. She must have been a courageous woman since she took mortgages on all three ships. The first married woman shipowner in Newfoundland was Caroline Winsor of Aquaforte; she held 13 shares in the *Pelter* when it was registered in 1844, i.e., 32 years before the Married Women's Property Act became law in the colony.

Use of the titles "Widow," "Spinster" and "Married Woman" persist throughout the period under study, though there are some interesting variations on the themes. Stated occupations are: Single Woman, Single Lady, Unmarried Woman, Married Lady, and in the later periods, Housewife and Homemaker. Also used are related terms such as Lady, Dame, Madame, Matron and Gentlewoman; the latter are generally indicative of high status.

Some of the other occupations listed for women in the ship registers are: Merchant, Trader, Dealer, Shopkeeper, Planter, Fisherman, Brewer, Accountant, Business Manager, Businessperson, Company Director, Clerk/Manager, Teacher, School Teacher, Religious Order, Secretary, Stenographer, Nurse, Seamstress, Milliner and Automotorcycle Dealer. The designation of women as "Mariner," but not "Master Mariner," is not uncommon; however, I suspect the former term in some areas was taken to be roughly equivalent to "Fisherman" (a term that is used occasionally for women) in other areas, but it may also have been used to designate women who actually went to sea. It is curious that the Quebec woman who had been designated "Manager" but whom the registrar decided should be designated as "Ship's Husband" was not instead called a "Managing Owner" since that term had long been in vogue. A woman in New Brunswick, Annie Seely, was listed as a "Tradesman – Non-marine"; her husband was a shopkeeper. Though the title "Agent" is not used, that appears to be the effect of designating Louisa Charlotte Boucherville (*neuve* Collar) in 1788 in Quebec as "subscribing" for the heirs of Joseph Collar, who are "non-subscribing." She was probably representing a non-resident family or possibly minors who were the inheritors.

The fact that women were, indeed, fishermen is demonstrated by a delightful vignette which Cyril Harris describes in his book *Northern Exposure: A Nova Scotia Boyhood.* He writes of the visit to Nova Scotia in 1904 of an American cricket team which was challenged to a game by the locals from Sandy Cove. The smartly dressed American team appeared on the field and looked with disdain on the local fishermen dressed in their overalls and denim shirts. The bowler for the locals turned out to be a young woman, though she bowled so well and carried herself so smartly that the visitors thought for a while that she was a young man. They discovered later that she was a lobster fisherman and that she worked in the boat with her father and sister every day. The "lobster girl" was responsible that day for the defeat of the fine Weymouth cricket team!

Exceptional, no doubt, is the story of Mildred Jewett of Nantucket, known locally as Madaket Millie, who was as strong at 11 years of age as any two Coast Guardsmen. Edward Rowe Snow, in his book *Women of the Sea*, writes that, as a mature woman, Millie "could roll over a dory, launch it, fish with the best of them, and could handle any sort of craft in almost any type of blow." Besting even this story is that of Capt. Linda Greenlaw of Topsham, Maine, who was the only woman swordfish captain in the Grand Banks fleet, and by 1997, had been fishing for 17 of her 36 years. Fishing out of Gloucester, Massachusetts, on the lunar cycle, she would return to port every 30 days with full loads of fish. In his book *The Perfect Storm*, Sebastian Junger characterized Greenlaw as "one of the best captains, period, on the entire East Coast," as he described the Halloween Gale of 1991. Greenlaw was a graduate of Colby College and was skipper of the swordfish boat *Hannah Boden*.

There are other interesting occupations of women designated in the early records. For example, when Eleanor Quinn of Rider's Harbour, Random Island, Newfoundland, married James Sullivan of Youghall, Ireland, in 1804, her occupation was given as "Boatkeeper"; that would be similar to ship's husband and probably meant that she was left in charge of fishing boats when the British owners returned home for the winter. Marjorie Clark of Bay Bulls was a boatkeeper sometime in the latter part of the eighteenth century. Hannah Langher in 1805 and Lydia Fifield in 1808, both of Trinity, gave the same occupation, boatkeeper, at the time of marriage. In 1815 Mary Gillet of Trinity, widow, married Charles Randell and her occupation is given as "Blacksmith." Perhaps she had inherited her late husband's business of blacksmithing and had carried on that business from the time of his death to the time of her remarriage. Madge ("Meg") Curlew was a blacksmith in Coley's Point, Newfoundland, through the 1920s; she was a big, raw-boned woman with large hands and wore a long leather apron at her work. She became a blacksmith by accident; the big wagon was broken, there was no one to fix it, and she did it, and that became her occupation.

As I have discussed these issues with people over the past several years, many have suggested snippets of information about women's unusual occupations in connection with the sea. Hannah Eliza Evans, whose husband was Edward, ran a supply store at Northern Arm for the George J. Carter fishing business of Herring Neck from about 1900 to 1914. Carter had bought the Edwin J. Duder business after it went bankrupt in the Bank Crash of 1894. Mary Olivia Noseworthy, whose husband had been Lorenzo, operated the family business at Spaniard's Bay, with sub-operations in Labrador, for three years from 1931 to 1933, and then phased out the business and moved to

St. John's. Thomas Collett, a merchant at Harbour Buffett in the 1870s may have left his estate, including ships, to Frances Bendle. A letter exists naming her children as beneficiaries and referring to her son Henry "carrying on the business, his mother not objecting." Other interesting variations on occupations exist. Cindy Parmiter of Point Leamington, whose father is Capt. Wes Parmiter, was a working crew member on her father's sealing ship in the 1970s. Sarah May Dempster, daughter of Jim Coombs of Plum Point, carried on her father's job of wharfinger (owner/manager of a wharf) until the late 1970s when CN Marine phased out its operations at Plum Point. Vera Abbott of Bonavista was a fisherperson and co-owner of boats and property with her husband Cecil for almost 20 years in the 1980s and 1990s, until her husband was tragically killed in a traffic accident.

Many women had a natural affinity for navigation and practiced the science of navigation aboard ship for many years. What is not as well known is that navigation was taught in schools in many parts of New England and in other places along the Atlantic coast. Navigation was offered as a subject in the "female department" of Dukes County Academy on Martha's Vineyard in the 1830s. The curriculum for girls included drawing, painting, projection and colouring maps. One of these New England women established a school of navigation in London, England. Miss Eliza Frame, a school teacher in Maitland, Nova Scotia, ran a quite remarkable school of navigation in that town. Donal Baird in his book *Women at Sea in the Age of Sail* tells the remarkable stories of Janey Congden Crowe, Helen (Smith) Grant and Bessie Hall, all Nova Scotia women who became successful navigators. There was said to be a female teacher in Bonavista, Newfoundland, who taught navigation in the late eighteenth century. Mrs. Makison, who had a dame school at Brigus, Newfoundland, may also have taught navigation. Miss Crib at King's Cove is also alleged to have taught navigation in school. It is known also that navigation was taught in the Roman Catholic school at Bay Bulls in the early 1840s. Memorizing the points of the compass was routinely practiced in many Newfoundland schools up until at least the late 1930s. I actually experienced that exhilarating pasttime in my early years in school. From at least 1917, F. J. Doyle operated a Nautical School in St. John's, and Capt. English, who was the Harbour Master, was one of his instructors. Before that time, it seems, examinations for Mates and Masters were conducted by the Board of Trade. There was a St. John's technical school and a local examining board which conducted examinations and issued Mate's and Master's Tickets (certificates of competency) as well as engineer's certificates of competency. The Nautical School appears to have taken over from the Board of Trade and also issued St. John's ambulance certificates. In October 1920 fifteen men (no women) were sitting for examinations at the Nautical School, and Drs. Carnell and Murphy were supervising the first-aid examinations.

In addition to being navigators, women were also shipbuilders. Perhaps the most prominent of the shipbuilders was Georgina Macdonald of Prince Edward Island who built the 179-ton ship *Corona* in 1904. Soon after the launching, she sold 24 of the 64 shares to John A. Macdonald and Robert N. Anderson, and two years later she sold the ship.

A quite different relationship with the sea was established by Miss Ivy Wamboldt, a ward of Captain Archie Publicover of La Have, Nova Scotia. In the late 1920s, the *E. P. Theriault* came into possession of Captain Publicover. This three-masted vessel had been built at Belliveau Cove, Nova Scotia, in 1919, was 310 tons net and had cost $62,000 in total. When he acquired the ship, Captain Publicover registered her in Ivy Wamboldt's name. Ivy held a Master's Certificate which she had acquired in the West Indies. For a time at least, Captain Publicover served as her mate, but Ivy was owner and master. Captain Wamboldt's many voyages from Nova Scotia to the West Indies during the 1930s are recorded in the newspapers of the day.

All these were women of strong character, shaped in part by the elements that surrounded them and bringing to their tasks an inner strength and grit and nerve that saw them bravely through anything that nature could put in their way.

CHAPTER THREE

Newfoundland Women and their Ships

In the early years, Newfoundland was like a giant fishing station moored permanently in the western Atlantic Ocean, and several European nations used it to pursue a migratory fishery, arriving in the spring with crews and supplies, fishing and curing fish through the summer and sending collector ships back to Europe with the finished product through the late summer and early fall. The merchants of Bristol may have discovered the rich fishing grounds of Newfoundland as early as 1480, and for almost one hundred years fishermen from the western coasts of Europe were practically undisturbed in their pursuit of the fishery here. Settlement in the New Found Land was actively discouraged by the government of the day. One small concession was made, however, in that a small crew was sometimes left to winter on the island in order to protect and expand their fishing "rooms," cut timber for wharves and fishing stages, build and repair small boats for the inshore fishery, and trap fur-bearing animals.

The Seventeenth Century

There were a few sporadic attempts by the British to colonize Newfoundland in the early years but these were limited to small groups of merchants. John Guy, on behalf of the London and

Bristol Company, visited Newfoundland in 1608 and settled at Cupids in 1610 with a group of 39 settlers, thus becoming the first governor of Newfoundland. In 1612 he brought more settlers. Women were among these early pioneers. John Mason succeeded Guy in 1616 as the second governor of Newfoundland. He and his wife Ann spent about four years on the island. Ann is credited with being the first teacher in Newfoundland because she successfully taught an American Aboriginal to speak English. John and Ann Mason jointly authored a highly accurate description of the island in a book entitled *A Briefe Discourse of the New-found-land.* George Calvert, the first Lord Baltimore, established a colony at Ferryland in 1621 and appointed Edward Wynne as governor of the colony. In the second group of settlers which came out in 1622, there were seven women. But Wynne, in a letter to Calvert, asked for more, especially for "a couple of strong maids, that (besides other worke) can both brew and bake." Contrasting with Edward Wynne's statement that "women would be necessary heere" was Captain Wheler's remark in the 1660s that "soe longe as there comes noe women they are not fixed." The latter statement reflected a battle that was raging in England about whether Newfoundland should be settled at all. The seasonal migratory fishery had been pretty well restricted to men only, and there were many strong West Country advocates of the status quo. At the same time that settlement was being consciously retarded in Newfoundland, an advocate for settlement in the American colonies was writing: "A plantation can never flourish till families be planted with the respects of Wives and children fix the people on the soyle." The presence of women was synonymous with settlement and community stability.

The formal colonies continued somewhat tentatively in Newfoundland. Lord Baltimore's colony at Ferryland was enlarged and became the Province of Avalon in 1623. Edward Wynne was replaced by Arthur Aston as governor in 1625 and Lord Baltimore finally came to Ferryland in 1627. The next year he brought his wife to Avalon. So dreadful and interminably long was the winter that it is believed that Lady Baltimore persuaded her husband to abandon the venture for something in the warm south. He applied for Maryland but died before the grant was awarded in 1632. The charter for the Province of Avalon was assumed by Cecil Calvert, the second Lord Baltimore, George's son, but he became preoccupied with the venture at Maryland and ruled Avalon through a series of governors until about 1637.

In that same year Sir David Kirke was awarded a charter for the entire island of Newfoundland, and he became governor of the colony on behalf of the Company of Adventurers to Newfoundland. In 1638 he arrived in Ferryland with about 100 colonists. His tenure was marked by the fiercest possible controversy with both French settlers and West Country merchants. As a result he was recalled to England in 1651 and placed in prison. Though he was eventually released from prison, it is highly unlikely that he ever returned to Ferryland. However, his wife Lady Sara Kirke did return and carried on the family fishing business for several years after her husband's death.

In Lady Sara Kirke we meet the first of our women shipowners in the Atlantic region. One of the informational history boards at present-day Ferryland states: "She rightfully deserves the title of British North America's first woman entrepreneur."

It is impossible to determine in what year Lady Kirke became a plantation owner or shipowner. She had first arrived in Ferryland in 1638. As the wife of the governor it is unlikely that she was having a "hands-on" involvement in the fishery in the early years. However, since she was carrying on an active fishing and supply venture according to the census of 1675, it may be safely concluded that she was a plantation owner, a shipowner and a boatowner from about 1643, if not before. She was not the only woman at Ferryland. Her sister, Lady Frances Hopkins, a political refugee whose husband was William Hopkins of Newport, moved to Ferryland in 1643 with the support and commendation of King Charles I. Lady Hopkins also appears in the 1675 census as a plantation owner pursuing an active fishery. Among the adults at Ferryland at the time of the census were Lady Kirke's four sons: George (and wife), Philip, David II, and Jarvase (or Jarvis).

They were all separate plantation owners. In Renooze (Renews) there was a cousin, John Kirke, and in Caplin Bay a Jer. Kirke, possibly Jarvis, with a second plantation. The Kirkes were a durable family, due probably in no small part to the Lady.

Their fortunes continued to be beset with troubles. Lady Kirke was replaced as first lady by Mrs. John Treworgie when the latter's husband was appointed governor in 1653. Lord Baltimore then contested the charter given to the Kirkes in 1637, though his counter claim was not recognized until 1660. Lady Hopkins went to England in that same year to present a petition to King Charles II concerning affairs in the island with specific reference to encroachments by the French and their continual military threat to English plantations. Lady Kirke took advantage of her sister's visit to the king to present a petition, signed by the principal inhabitants of the area, "praying" that her son George be made governor of Newfoundland. Coincidentally, Governor John Treworgie had just completed his term. No action was taken on the latter recommendation. Finally, in 1661, Lady Kirke was forced to surrender title to all property claimed by Lord Baltimore. Though she continued to live in the house that her husband had built, she was required to pay rent for it to Baltimore's agents. Lady Hopkins returned once more to England to protest against the treatment of the Kirkes by the government of the day, but without success.

From this historical record it may be determined both that these were strong-willed, determined women who thrived in adversity and that the fishing and supply venture they were pursuing was a very lucrative business.

Sir John Berry's census of 1675 covered the east coast of Newfoundland, from Trepassey in the south to Bonavista in the north. The census records for Ferryland indicate that Lady Kirke had sixteen male servants, one female servant, one dwelling house, one lodgings house, four boats, one stage (for "making"/curing fish), one storehouse, four rooms (i.e., four large separate spaces on or near the shore for fishery work), one train vat (fish oil), and that she caught the allowable quota for the year, 150 quintals of fish (a quintal is 112 pounds). Her sister, Lady Hopkins, employed thirteen male servants, one female servant, had one dwelling house, two lodgings houses, three boats, one stage, two storehouses, one train vat, three rooms, and had also caught 150 quintals of fish. George and David Kirke II had plantations that were slightly larger than their mother's and aunt's.

It can be concluded from census evidence and other historical data that Lady Kirke operated her fishing business for approximately 40 years and that Lady Hopkins had operated hers for 30-plus years. From the Berry census it can be concluded that women in several communities were boat-owners from at least the mid-1600s. Boats would have been used to carry on an inshore fishery. Boats are generally distinguished from ships in that the latter are "of considerable size," and boats are generally open vessels, i.e., undecked. If the Kirkes owned ships (as seems likely from the number of servants they employed), these would have been registered in a British port since there was no registration of ships in Newfoundland at this early period. Ships built by English planters or merchants in Newfoundland during the seventeenth century would have been registered in a British port. In any event, if the Kirkes owned ships, these would have been anchored in the Ferryland harbour or tied to a wharf and not used to pursue an inshore fishery. Since the Kirkes were now established settlers they probably sold their fish and oil to collector ships operating out of England.

Other women were arriving in Newfoundland in the early 1600s; for example, Elizabeth Matthews was in Bay Bulls in 1641 and Margery Burt was in the same place in 1651. Mrs. Gilder was involved in the fishery business in 1663, place unknown. There were at least nine other women boat-owners listed in the 1675 census in six other communities between Bonavista and Trepassey: Joan Clay at Bay Roberts; Margaret Taverner at Bay Verds (Bay de Verde); Sarah Edwards at Tarr Bay (Torbay); Emma Horton, Barbara Horton and Joan Hibbs at Harbour Grace; and Widow Long (or Lony), Widow Sertall and Mary Furze at St. John's. All of the women are clearly designated as widows except Margaret Taverner of Bay Verds. She was in fact a married woman who resided in Bay de Verde and operated a fishing room there because her husband William was ill at home in Poole. Her sons Andrew and Robert operated fishing establishments there also. The Taverners had been in Bay de Verde from the 1650s. Other women reputed to have been involved in the fishery business were: Emelin Garland in 1675; Widows Wood, Holmen and Gresham in 1676; Lucy Guy and Widow Hanan in 1677; and Widow Roberts in 1681. Also in 1681 there is mention of Widow Bearns living in St. John's with two children, two boats and ten servants. The English population along the east coast in 1675 totaled 1,367 of which 114 were planters. If women in charge of fishing plantations are counted among the planters, they represent about ten percent of the group, which is quite remarkable. The censuses of 1675 and 1681 demonstrate also that widows employed an average of thirteen servants on their plantations compared with an average of nine for all plantations.

A petition was sent from Ferryland to King William III on November 27, 1696, describing continuing French aggression against the settlers, the destruction of 12,000 pounds sterling in property, and the banishment by the French of the entire population of 150 persons. The petition was signed by 40 men. Lady Kirke and Lady Hopkins had long since died or departed; George Kirke had moved to St. John's sometime after the 1675 census; David Kirke II was imprisoned by the French at Placentia, where he died in 1696 or 1697. Following the destruction of several English settlements by the French in 1697, there was at least a tacit tolerance by the British of their own settlers along the east coast and some modest efforts made to supply a military presence for defense purposes. When the British government passed the Act to Encourage the Trade to Newfoundland in 1699, its main thrust was still to promote the existing migratory fishery, but it also included the right to property which had been held in 1685 or before.

After David Kirke II died in 1696 or 1697, his widow Mary married the St. John's merchant James Benger in 1699; she then claimed the Ferryland plantation, and they established a business there which they operated for many years. Mary Kirke's decision to remarry was one option for a widow, and it was an option that was taken by many widows, though widows were not as likely to remarry as widowers. Gordon Handcock writes in his book *Soe longe as there comes noe women: Origins of English Settlement in Newfoundland:* "The (Mary) Kirke example does illustrate at least several social processes important for permanency. These include the transfer of property through females, the social situation of the widowed female who found herself stranded in Newfoundland and enforced by circumstances to seek through early remarriage some measure of security and, most importantly, the social interaction between the settled population and the migrating labour force."

The main reason for remarriage was economic; it was clear that husband and wife working together formed a strong economic unit. However, many widows were not prepared to surrender the independent status of being head of a household, particularly if the wife and husband had already built up a viable economic unit which she understood and was nearly in control of by virtue of the fact that she almost certainly had more schooling than her husband and had probably kept the accounts, handled the money and been involved in marketing the product. Peter Pope in his thesis *The South Avalon Planters 1630 to 1700* writes: "By becoming planters themselves, Newfoundland widows were adopting a male role, but this was not as anomalous as it may seem, within the context of a patriarchal society. Women's economic functions were not completely distinguished from men's in Newfoundland any more than they were in pre-industrial Europe; furthermore English women had an acknowledged right to assume male roles under certain circumstances."

Women in Newfoundland during this period functioned in a male-dominated, patriarchal society, but those who were involved with plantations enjoyed a higher status than housewife. To quote Peter Pope again: "Newfoundland planters' wives participated at least as fully in economic power as women anywhere in the seventeenth century. In other words, they were powerful relative to their sisters elsewhere, in a century in which women were powerful, relative to their great-granddaughters." These women were not confined to shore activities but were involved in more substantive duties such as curing, processing and marketing fish. Yet these duties would not have excused them from the normal female activities of the day such as baking, brewing, and the care of livestock and kitchen gardens. Depending on the size of the business there were likely female servants to assist with these duties, but the wife or widow would have been an overseer. A wife, in fact, was expected, according to English custom, to act in her husband's place if he became ill or was away at sea.

The Eighteenth Century

From court records and other sources it can be demonstrated that women continued to hold shoreline property (plantations or "rooms") and pursue an inshore fishery with small boats and hired servants.

In 1701 Thomas and Margaret Lamb were planters in Little Placentia and Great Placentia. In that same year Margaret had brought an action against Simon Honeyburn, agent for Messrs. Joseph and Moses Neave, merchants of Poole, Dorset, and she posted a bond of 40 pounds sterling. Honeyburn was ordered to keep the peace for one year and appear in the next Sessions Court to answer the complaint. (Since this was given as a note in the 1757 court records it is not known how the matter was settled.) In 1757 Margaret, who was now residing in "the City of Waterford in the Kingdom of Ireland," bequeathed her property through the court to James and Thomas Lamb, children of William and Mary Lamb of Great Placentia. The bequest reads in part: "I do give and grant…the plantation with all the appurtenances thereunto belonging situate and lying and being in the southward part of Great Placentia aforesaid in as large ample and beneficial a manner as the same was conveyed to me by virtue of a Bill of Sale from Mrs. Jane Grossard of Great Placentia…" This is a clear case of property transfer by bill of sale between one woman and another. It may also be inferred from this document that Margaret and her husband operated separate plantations in the two communities. Ann Lake of Placentia, in the late 1700s, petitioned the court after she was beaten by six of *her* male servants. Her husband George was a planter in nearby Paradise. The court found in favour of Ann Lake and assessed fines to four of the six. Thomas *and* Catherine (Boudet) Ouldin sold *their* property in Great Placentia for 30 pounds sterling to Richard Welsh, merchant. The record clearly states that the property belonged to *both* Thomas and Catherine. Ann Redmond of Placentia took John Kavanaugh to court in 1753 because she had loaned him a boat and it was in damaged condition when it was returned to her; Kavanaugh was ordered to repair the boat. In 1786 Thomas Roach, planter, purchased a plantation from Mrs. Hogan who was the widow of William Hogan, planter at Placentia.

The situation is similar in the court records for Ferryland. There is a most intriguing case in 1758 in which William Ludevig and William Keane, Esq., request the court to divide the property of Mrs. Mary Ludevig, deceased, between her daughter and her son. A party, appointed by the court, met on the premises and made an equal division of the property; the party consisted of John Benger (undoubtedly a son of James Benger whom Mary Kirke had married), Robert Carter, John Walley, J. Carter and Peter Weston. The son is William Ludevig; the daughter must have been a single woman and bears the same name as her mother, Mary Ludevig. The division of property was made in such a way that each of them could carry on an independent fishery, apart from sharing certain facilities. Mary is given the north shore and the bread cott; William is given the south shore by itself. The wharf and the wharf shore are to be jointly used by sister and brother. Mary is then given Mr. Bowles' house and the hen house, workroom and the house where Doctor Ball lived formerly and the meadow and garden belonging to this house. William is given Doctor Moor's house with its gardens, woodhouse, the meadow, Collins' house and garden. Mary is then given the north stage and the vacant spot to build another. William is given the south stage and the leaney (linney) stage. Mary is then given the lower room, and Fanny's cookroom.

William is given the upper room, Wm. Dyer's cookroom, with the upper park. The flake room, the beach, and the breakas (ers?) are to be divided equally between sister and brother. This division of property occurred in 1754 but was not entered into the court records until 1758. There is a reference in the court records of 1773 to "Mr. Ludwick's Plantation," so it is obvious that the family was still there. In his will Peter Weston refers to William Ludevig as his nephew. It is interesting that such an equal division of property is made between son and daughter by five males. It clearly demonstrates the prevailing British practice.

In that same year, 1758, Joseph Seal gave Honea Hunt, whose husband was William Hunt, a house and piece of land in Ferryland Harbour, and the court states that "all…are to give Honea Hunt quiet and peaceable possession of the same." No mention is made of the reason, or the possible relation between Seal and Hunt, or joint ownership with the husband. Mary Shea in 1773 applied for the use of two small plots of land adjacent to her house on the north-west side of Ferryland, land which had been in her possession for a number of years, and she is instructed to use the land "so long as you shall employ it to the advantage of the Fishery," and if the governor agrees.

Catherine Clements, widow, was at the centre of a case in 1785 in a dispute over waterfront property. Mrs. Clements was the owner of two boats and a skiff (a small, light fishing boat); she had formerly owned four boats. She probably now had 15 to 20 servants, several of whom testified on her behalf in court. James Rouse had built flakes on disputed ground which Mrs. Clements had used for several years but which she did not claim. Nevertheless she and three of her men pulled down the flakes because Rouse and his men had prevented her from washing her fish in the landwash. She also claimed to have an order to do so from the Fishing Admiral at Renews. The plantation which Mrs. Clements owned had been known 23 years earlier as the Potsbury Plantation, and it had been owned after that by John and David Miland who sold it to the Clementses. Justice Carter confirmed this sale, stating that the property had once belonged to his father.

Unfortunately the court record ends in mid-sentence with the words: "...but as the said Clements had dried her fish there before they had given it her, she being last in possession..." which seems to imply that a decision was made in favour of the feisty Mrs. Clements.

A few women at Bay Bulls are known to have owned plantations in the latter part of the eighteenth century. One of these was Betty Pottery of Sidmouth, England. She owned a fishing room in Bay Bulls by at least 1768, but it is not known for how long. Marjorie Clark also owned property and a number of houses in the town.

Farther north along the east coast of Newfoundland the situation was similar. At Bricus (now Brigus) in the 1780s, Edmond Dunfey lodged a complaint in court against James Connoly because he entered without permission on land *which had been in possession of his wife, Austis Dunfey*, for at least 14 or 15 years, and cutting the grass and carrying off the hay. Connoly was ordered to restore the hay and pay five shillings. Note that it is *the wife*, not the husband, who owns the separate property.

There are some interesting cases described in Warwick Smith's manuscript on Trinity, covering the 1700s. Though the total amount of her indebtedness is not revealed, Mrs. Mary Waterman was ordered to pay her creditors annual amounts as follows until her debts were discharged: to Thomas Webb and Isaac Lester, 15 pounds sterling each; to Messrs. Masters and Balard, 2 pounds; and to Joseph White, 1 pound. As long as she kept up her payments, "she was not to be subject to distraint and could go on with the fishery." After her death an inventory of her effects shows "casting nets, caplin seine, cod seine (and) lance seine." Her boats must by now have been disposed of, or were considered too large to be covered under the term "effects." In 1723 the widow of William Harvey had conveyed to her son-in-law, Philip Sweate, by Deed of Gift and for consideration of 40 pounds sterling, property which had come into possession of her husband in 1699. In another case, William Jones was accused of cutting grass on the property of Mrs. Mary Sweet, widow, at Ant Cove. Wm. Jolliffe, through his will of 1759, left a plantation at Trinity Harbour to his daughter-in-law, Ann Jolliffe, and other plantations at Bay de Verde and Fogo to his sons James and Peter.

References in various manuscript collections confirm women's involvement in ownership and disposition of property throughout the eighteenth century. In a bill of sale dated 1711 at Old Perlican, Mary Carter, widow and "relict" of John Carter, a deceased planter from the Towne and Country of Poole, with her eldest son Roger Carter and John Carter, the younger, disposed of *their* plantation on the north side of the harbour for 20 pounds sterling by transferring ownership to John Barrett. When Anthony Varder retired to England in 1714, he left a plantation at Bay Roberts to his daughter, Ann Rock, and a second plantation at Port de Grave to another daughter, Elizabeth Brady, whose husband was Jacob. In 1737 John Pelley, a planter at Fogo and a native of Winkton, Parish of Christchurch, Twineham, in the County of Southhampton, made his will and left his estate to his wife Mary and his two young children, Mary and James, including the profits from both boats and from a sawing and boatbuilding voyage which had just been sent abroad. He also left his holdings at Winkton to his wife Mary; in the event of her remarriage these holding would go to his daughter Mary, with 20 pounds sterling to his son James.

Catherine Thistle of Harbour Grace requested and got action from the governor against Jersey men who were encroaching on her plantation in 1757. Elizabeth Abbot of Bonavista complained of being defrauded of a plantation that had been her grandfather's. It had been claimed by Jonas Newell, who stated that he was the closest male relative and therefore had that right. The governor decided against Jonas Newell and in favour of Elizabeth Abbott. William Winsor of Fogo in 1784 to 1786 must have been bankrolled by a Mrs. Jeans of Poole since there is a note in the manuscript record that he made a payment to her in connection with shipping a cargo from Newfoundland. Nicholas Ellis of Woodbury, Devon, and his wife Anne (nee Smith, daughter of William Smith) became owners of property in St. John's in 1777 through the will of Anne's grandfather, Christopher Archer. The will states that Anne is his eldest granddaughter. The property is described as follows: "All those Messuages Tenements or Dwellinghouses Rooms Stages Flakes Lands Plantations and Hereditaments with their and every of their Appurtenances...lying and being in the Harbour of St. John's..." It is not known whether Anne or Nicholas ever visited or occupied this extensive property, but it is known that John Stripling, Esq., was their tenant in 1777. Sarah Richards, widow of Bristol, through her will of 1772, left property in Newfoundland to a number of her sons and daughters. Likewise, Joseph Bower, Jr., of Bristol, in 1793 left his plantations at Harbour Grace to his daughter, Susanna (Heighington) Roberts. Susanna's name appears in the *Conception Bay Plantation Book* as purchasing property on the north side of Musketta.

Gordon Handcock identifies an early woman shipowner in the person of Mary Luckem, a widow of St. Nicholas, England. She owned shares during the 1780s in the brig *Aurora*, along with Thomas Row, merchant, and Capt. William Fox, mariner of Shaldon. The Fox family had a connection with Newfoundland through Capt. Fox's father, who for some years had been a byeboatkeeper in the cod fishery.

Women also rented property for their own use and leased their own property for others to use. Two enterprising women of St. John's, Patience Pynn and Elizabeth Pippy, owned plantations in Musketta and leased them in 1765 and 1796. These and others like them were the early business managers. Several women owned more than one piece of property and leasing seems to have been their preferred option. Annual rents ranged from 13 shillings to 13 pounds sterling, and the periods of the leases lasted from one to fourteen years. Richard Burt and his wife Saranna had leased a plantation for their fishery at Old Perlican in 1717. They were originally from Winkton. They later bought this plantation and it was sold *by them* in 1728 to John Durrell for 70 pounds sterling. Alice Thomas of Ferryland in 1772 petitioned the governor to take back from Samuel Jago two boat rooms which had been owned by her deceased husband, Thomas Thomas, and shared with his partner Jago. She stated that she could get 15 pounds sterling for summer rental but that Jago had given permission to

James Shortall to spread his fish in the rooms. Alice stated that she was prepared to seize the fish. Both parties agreed to have the matter settled by Messrs. Sparks and Drew or by a third person, or forfeit 30 pounds sterling each. Anastatia Salmon, widow of George Salmon, rented property in Placentia on a three-year term. Widow Sarah St. Croix was listed as "the Proprietor of a store house" and she had rented it in 1801 to Larkin & Brennan for one year at 10 pounds sterling. It was stated in the court record that she now wished to terminate this lease in order to rent the storehouse to John Power. This was held to be a serious matter and it was referred to the Surrogate Court for a decision.

Aaron Thomas in his journal of 1794 mentions several women of status whom he had met in his travels. At Ferryland he met Mrs. Keane whose husband had come from England about 55 years previously as an "Adventurer" or "green man," that is, with no experience as a fisherman, and had, through hard work, become "the most affluent Fisherman in Ferryland." Mrs. Keane was now a widow, lived in a place called The Grove and "was in good circumstances with 14 cows." Thomas enjoyed her hospitality, but even more the treatment he received from Widow Tree, who lived in "a House of Entertainment" called the London Inn. Her husband, Captain Tree, had been an American Loyalist driven out of Boston in 1783, losing all his property there. Mrs. Tree was then about 50, had "a large and roomy House and Genteel furniture" and served him a meal "on Four Covers." He remarked: "Had an Epicure been one of the Guests he could not a found fault with a single Dish." Thomas mentions two other widows, Mrs. Harty of Portugal Cove and Mrs. Skinner.

Though the *Conception Bay Plantation Book* was compiled in 1805, most of the transactions have to do with the 1700s so it seems more fitting to consider it and the other plantation books under the eighteenth century rather than the nineteenth. See Appendix B for more details on these plantations or rooms.

There are about 147 women named as owners of about 156 plantations or referred to in bequests to their children or grandchildren or other relatives in the *Conception Bay Plantation Book*. Depending on how you figure it, about 13% of the owners were women, or about 15% of the plantations were either owned by women or bequeathed by them to existing owners. As a measure of involvement in the economic activity of the colony, the large percentage of plantations owned by women is quite remarkable.

Women came into possession of these plantations by various means: by purchase, by inheritance, by deed of gift, by exchange, with permission of the governor, by claim of its lying void, by cutting it out of the woods with permission of the king, and so on. Most of the purchases were by amounts that ranged between 20 pounds and 40 pounds sterling, a few were bought for about 90 pounds, and there were at least three that were bought for amounts ranging between 120 and 150 pounds.

Inheritance was a common way to acquire property, as would be expected. Women inherited from husbands, mothers, fathers, grandmothers, grandfathers, fathers-in-law, brothers-in-law, uncles, and even mothers-in-law. Apart from legal inheritance, property was also acquired by deed of gift from family members. It is not surprising that, because of inheritance through the patriarchal line, 16 cases were found in which men received property from their fathers-in-law or their wives' uncles.

Two very enterprising women acquired land by exchange. While some pieces of property occasionally sold for very small amounts (e.g. 2 pounds or 5 pounds), one woman acquired property in 1774 in exchange for a pair of shoes, and another purchased property in Carbonear in 1787 for a whipsaw worth 45 shillings, a very astute deal indeed! Two women acquired their plantations by permission of the governor, and a large number of women acquired property that they had "cut and cleared agreeable to the Act of William III chap. 25 Sec. 7." Four women claimed property "in consequence of it lying void."

There were also several cases of joint ownership, usually husband and wife, e.g., John and Sarah Wells of Lower Island Cove;

Edward and Mary Merrigan of Collier's Dock; and Mary and William Balding of Harbour Grace. Mary's name appears first in this entry. However, there were other instances of joint ownership: John, James, Richard and Thomas English owned property together with the widow of William English at Devil's Cove; Mary Wheaton, James Cole, John Cole, Mary Cullin and Mary Cole in Colliers; William Mugford and two sisters (unnamed) at Port de Grave; Ann, Catherine and William Narrowcot at Colliers; Edward (or Edwina?) Nugent, Charity Noseworthy and Patience Dwyer at Port de Grave; William Antle and Widow Antle at Brigus; William, Jno., Peter and Ann (Toque) Howell at Carbonear (Ann Toque appears later as a shipowner); and the descendants of Joseph Martin and Widow Mary Martin at Harbour Grace. There were two cases in which women held a plantation in joint ownership: Catherine Bryan and Mary Lee of Colliers in 1793, and Mary and Ann Moore of Harbour Grace in 1805. In both cases they were leasing the property to others. In another instance Ann Wells was occupying property at Carbonear in 1786 which was owned jointly by her and G. & S. Kemp & Co.

There is a most interesting case at Port de Grave in which a husband and wife team held two distinct properties. Both properties had been in possession of the Anthony family for 105 years, i.e., since 1700. Matthew Anthony bequeathed one piece of property to his daughter Ann Coveduck and the other piece to his son-in-law John Coveduck, Ann's husband. Matthew's obvious intention was that his daughter and her husband would operate the two plantations in concert.

At Port de Grave, George Dawe, Sr., came into possession of a plantation through his mother's will in 1755, and the record notes that the plantation was "possessed by his ancestors for 160 years," i.e., since 1645 (taking 1805 as the date from which it should be calculated backwards, though earlier dates are given in other entries from which it may have been backdated). Several other properties are noted to have been in possession of certain families for terms that range from 30 to 150 years.

There are a few instances in which women had formalized the title of the family business, undoubtedly for the purpose of dealing with the merchant or supplier, e.g., Widow King & Sons of Brigus occupied a plantation which had been in their possession since 1784; Mary Dalton & Sons owned property at Broad Cove which she was renting to Morgan Hingay in 1777 for 10 shillings per annum (Mary also owned shoreline property at Western Bay in 1784 which consisted of one stage, one flake, three houses, two gardens and one meadow); Widow French owned "French's Plantation" at Harbour Grace; and the firm of Ruth Collins & Sons is mentioned in the court records of Placentia in 1802 and 1806 when Patrick and Dennis Murphy are ordered by the court to leave 12 feet of space between their new stage and the existing stage of Ruth Collins & Sons. Sean Cadigan, in his thesis, also mentions Elizabeth Webber & Sons at Harbour Grace, probably in the 1760s.

In his carefully researched book *The Settling of Spaniard's Bay*, Eric Martin Gosse makes a very strong argument for regarding the families listed in the 1805 *Plantation Book* as "founding families" of the area. He proves his point for Spaniard's Bay, and we can extrapolate his argument to cover Conception Bay in general. Many of the women listed in the *Conception Bay Plantation Book* operated pioneer fishing plantations and were significantly involved in the early economy of the colony. A great deal of weight must be given to Eric Gosse's arguments because of the perspective from which he speaks. After graduating from college, he was employed with the family firm of G. & M. Gosse and was manager of the firm both at Spaniard's Bay and Emily Harbour, Labrador. In 1937 he was employed with the Newfoundland Fisheries Board. After serving in World War II, he became a trade representative of the Fisheries Board, and later, the Canadian government, in the West Indies. In 1956 he was appointed Deputy Minister of Fisheries for Newfoundland and retained this position until he retired in 1973. I interviewed him at his home in Spaniard's Bay in 1994, and we corresponded for about two years after that date.

The *Register of Fishing Rooms in Bonavista Bay 1805-1806* reflects a similar picture to that of the *Conception Bay Plantation Book* but the numbers are much smaller. On Pond Island, at the eastern end of the harbour of Greenspond, was Mary Hutchin's room, which she had gained by inheritance and which had been built up originally by her family. Hannah Pladwell owned Pladwell's room at Bayley's Cove, Bonavista, by inheritance from her husband, and was renting it to William Pladwell. Sarah Abbott owned Stephen Abbott's room at Bayley's Cove and was renting it. Brown's room at Bonavista was owned by Jos. Brown & Co. and the brothers and sisters, and John Ward was leasing it. The Browns also owned a second plantation in the town by right of inheritance.

The situation is similar in the *Register of Fishing Rooms, Twillingate and Places Adjacent, 1806.* Benjamin Brooks' room at Back Harbour, Twillingate, was acquired by marriage with the Widow Ridout. Joseph Thomas came into possession of his room by marrying the Widow Ware. William Banks married the Widow Bide and thus acquired a room at Twillingate. Widow Smith at Riverhead, Twillingate, had inherited her room from her late husband, and then married John Vincent so the room was listed in his name in 1806. Mrs. Parsons' room at Jenkins Cove, Twillingate, was occupied by Widow Parsons and she, apparently, was carrying on a fishing venture.

There are also references to "Independent" women in the John Slade & Co. ledgers covering the period 1783 to 1792. In an unpublished article in 2001, Don Bennett described "Independent" as "independent schooner men, often with Slade financing." At least two women are so characterized in the Slade ledgers. Susannah Thomes, widow of Jacob, who was a servant of Slade from 1783 to 1789, is listed as "Independent" from 1789 to 1792. John Symes & Co. was an associate of Slade from 1783 to 1788, "has a separate English account, used to bolster the Newfoundland venture"; John apparently died in 1788, and Mary, his widow, "took over management until liquidation" in 1789. In 1789, William Towler appears as a representative for Mary Symes, obviously her attorney. Hanna Davis was a boarding house keeper, probably in Fogo, from 1787 to 1791. Dorothy Burton did "doctoring" in her home from 1782 to 1786, until Dr. Stotesbury arrived. And Mrs. John Primer, wife of a planter, picked 84 gallons of berries for Slade in 1791 and 127 gallons in 1792. That would be in addition to other duties at their fishing room. From such planters Slade collected furs, salmon, seals, codfish and berries to trade at Poole, Waterford, Ireland, and in Spain, Portugal and Italy.

The Nineteenth and Twentieth Centuries

It is not until the beginning of the nineteenth century that we can discover the specifics about women owning ships or shares in ships. It was very late in the eighteenth century or early in the nineteenth century before registrars of shipping were appointed in what are now the Atlantic provinces of Canada and Quebec. Before this, ships were registered in European ports. Before we look at the ship registers in the New World, there are a few references early in the nineteenth century that are worthy of our attention.

First, there is the remarkable story of Mrs. Anne Huelen (or Hulan), a farmer/trader/shipowner in the Codroy Valley of Newfoundland's west coast. It is hard to know where to place Mrs. Huelen because she seems to have lived from about 1740 to 1840. She had two distinguished visitors, in 1822 and 1835, and it is from them that we have information about her. She remembered Capt. James Cook who was doing survey work for Britain on the Newfoundland coasts from 1763 to 1767. William Epps Cormack visited her as he completed his walk across Newfoundland in 1822, the first white man to do so. Cormack comments: "…the farm of my hostess Mrs. Hulan at the second Barachois River…her stock consists of 6 milch cows, beside other cattle. The dairy could not be surpassed for neatness and cleanliness, and the butter and cheese were excellent. Butter…was sold, part to residents of other places in the Bay and part to trading vessels coming to the coast in summer. The cellar was full of potatoes and other vegetables for winter use. She was also an experimental farmer, and exhibited eight different kinds of potatoes, all possessing different qualities to recommend them…Of domestic poultry there was an ample stock. Mrs. Hulan, although not a native, had lived in St. George's Bay upwards of 60 years...She is indefatigably industrious and useful, and immediately or remotely to, or connected with, the whole population of the bay, over whom she commands a remarkable degree of material influence and respect."

Mrs. Huelen's second distinguished visitor was Archdeacon Edward Wix, a Church of England missionary from Great Britain, who was making a visit by boat to communities of the southwest and west coast of the island. On June 6, 1835, he records in his diary that Mrs. Huelen (how he spells her name) was "a cheerful old lady," "the mother of the settlement" and "a native" of the country. He then records an amazing experience that Mrs. Huelen had in 1814. She was then a widow and she and her daughter and the crew were taking her schooner with a load of cured salmon to St. John's and for the "arrangement of her affairs," when the ship was captured by an American privateer and taken to New York. Her cargo was confiscated and sold there under a writ of "Venditioni expanos," and her "pass papers" were signed by none other than James Monroe, Secretary to the President of the United States, who would three years later become president. A number of benevolent Americans in New York came to her aid, especially Sophia and Elihu Doty, and helped her provision her ship for the return trip to Newfoundland. On the way back she stopped at St. John's to put her business affairs in order. Out of gratitude she later arranged to have two of her grandchildren named after Mr. and Mrs. Doty. Don Morris, writing about Mrs. Huelen many years later remarked: "The lady had no renown in the U.S., but on the west coast of this island she was hailed as a super active, successful trader, a sparkplug of a woman who got things done."

Widows who were apparently carrying on a business are identified in the manuscript *Number of Inhabitants in the Harbours of Brigus, Cupids, Bareneed, Port de Grave for 1817*. The fact that these widows had male servants, female servants, were supplied by merchants, and are designated as being "well off" or "very well off" may be taken as proof that they were in business. These are: Mrs. Noreman and Mrs. Best of Brigus; Mrs. Hussey, Mrs. Furneaux,

Mrs. Grenley and Mrs. Kenedy of Bareneed/Port de Grave. There are ten widows listed for these latter two places and Cupids who are designated as "distressed." An unpublished manuscript entitled *Fishermen of Fogo, Twillingate and Change Islands*, compiled by Angus and Verna Elliott from "various sources," purports to list several women who were involved in early fishing ventures. The version I saw had a listing only for the A's and B's and included the following names: Dinah Adams, Joe Batt's Arm, 1821; Ann Brown, Joe Batt's Arm, 1821; Maria Burges, Twillingate, 1821; Mary Ann Braseel, Fogo, 1823 (Mi'kmaq); Mary (Kilkenny) Aylward, Ragged Harbour, married, 1829; Mary Brennan, Joe Batt's Arm, 1831; Rachel Brown, Herring Neck, 1857; Mary Boone, Twillingate, 1870; Julia Ann Banks, Back Cove, Fogo, 1879; Elizabeth Barnes, Fogo, born 1866, married 1882; Maria Blandford, Fogo, 1892. I cannot vouch for the authenticity of these records.

The lucrative salmon fishery on the Campbellton River had attracted the firm of Garland, Read & Co. at least by the early 1800s, and in 1823 they sold these rights to John Genge. Genge sold these to Joseph Hornett, and during the period 1861 to 1875, it was stated that "the river is now claimed and occupied by a widow of that name." This may have been Ann Hornett or possibly Sarah Hornett who appeared in the church records for Change Islands with her husband Joseph for the baptism of their daughter Nancy in 1821. Her son William Hornett would eventually take over the salmon business on the Campbellton River.

An interesting instance of joint ownership occurred at Trinity where Joseph Purchase and his wife Amelia (nee Newell) sold their plantation in 1848 to Patrick Fowlow. In 1837 Thomas Newell, Sr., had bequeathed the plantation to his daughter Amelia, and it consisted of "a certain Room or Plantation called Newell's Point at the entrance of Gat's Cove in the N. W. Arm of Trinity." The bill of sale to Patrick Fowlow reads: "Know all men that I Joseph Purchase Husband of Amelia Purchase she consenting thereto, as witness her signature to these presents, for and in consideration of the sum of 23 pounds currency...sell to Patrick Fowlow planter and dealer of Trinity..." The plantation which was Amelia's was vested in the name of her husband, but it is interesting to note the references to her "consenting" and to her "signature." One wonders if she had insisted on this, for she no doubt possessed the moral authority to do so. She probably kept the accounts and ran the business end of the family venture, for she signed her name, but her husband simply made his mark on the bill of sale. Later, in 1885, after Patrick Fowlow's death, land was granted to Widow Mary Fowlow and her sons John, Thomas, Philip and Martin, all of whom are called planters. In the Newfoundland Crown Lands Grant, the widow's name preceded those of her sons in the five times these occur. Three acres, three roods and thirty perchas were added to their holdings by this grant.

Canadian ship registers from the late-eighteenth century to the mid-twentieth century clearly indicate that widows, single women and married women owned ships and shares in ships on a fairly regular basis. Their numbers were never great, but they were a presence, and their involvement in the business of shipowning is worthy of the closest study. Their actions anticipated legal changes and, indeed, inculcated a climate of change that would influence future legal and social decisions.

The more than 400 women whose names appear as shipowners in the Newfoundland ship registers and other records were scattered all over the island in about 130 communities along the coast. By way of a rough summary, there were 80-plus women in St. John's, 15-plus on the Southern Shore of the Avalon Peninsula, 5 in Placentia Bay East, about 70 in Placentia Bay West and the Burin Peninsula, about 45 in Conception Bay, about 25 in Trinity Bay and the same in Bonavista Bay, 30-plus in Notre Dame Bay and White Bay, about 25 on the West Coast, 40-plus on the South-West Coast, about 30 on the South Central Coast, and one in Labrador. The largest concentrations were in St. John's (about 80), Grand Bank (20), Carbonear and Rose Blanche (17 each), and Burin (12). Twillingate and Harbour Grace had 8 women shipowners; Harbour Breton, Wesleyville and Harbour Le Cou had 6; Brigus, Fortune, Ramea and Belleoram had 5; most of the other communities had 1 or 2 women, and a few had 3 or 4.

Without naming every individual woman in the records, perhaps the best way to deal with the wealth of information about them is to describe a few representative women by ownership patterns, i.e., whether they were sole owners, joint owners, co-partners in trade, managing owners, purchasers of shares, whether they were mortgagees or took mortgages themselves, whether they inherited ships and sold them quickly or remained active in the business, and so forth. The main focus will be on the earlier period and up to the early 1920s, although I searched the records up to 1990.

The two favoured ownership patterns emerging in the records were sole ownership by a woman and joint ownership with men or other women. The term "sole ownership" as used in the ship registers usually means that there was one or two or three or more *exclusive* owners, but it is here taken to mean *exclusive ownership by one woman*. Again, by way of rough summary, there were more than 220 instances of women being sole owners of ships, about 150 instances of joint ownership (including about 40 cases in which a woman owned jointly with her husband and other men and about 10 cases in which a woman owned jointly with another woman), 3 cases in which women were designated "co-partners in trade," 2 cases in which men appointed women as managing owners, more than 30 cases in which women were designated managing owners, about 25 cases in which women appointed men as managing owners, more than 30 cases in which women were mortgagees, almost 50 cases in which women took mortgages on their ships, at least 13 cases in which women did not hold a mortgage but the man to whom they sold had to take one, 6 cases in which women sold their ships to other women, 2 cases in which couples sold directly to other couples, about 40 cases of husbands "selling" to wives and 5 cases of wives selling to husbands, about 80 cases in which a woman inherited a ship and sold it within one or two years, about 50 cases in which a woman inherited a ship and remained active in the business, almost 20 cases in which a woman apparently "staked" a planter to build her ship or hired a shipbuilder for that purpose, and at least 3 cases in which a woman's ship was engaged in the seal hunt. Of additional interest is the fact that there were only 2 instances in which a woman shipowner died with a will and over 40 cases in which the woman died intestate; also there were at least 21 cases in which a woman named her ship after herself, 13 hints of scandal in connection with a woman's ship, and 5 cases of minors inheriting shares in ships. There were almost 50 cases of a woman being appointed sole executrix of a will, about 25 cases in which a woman was appointed as an executrix along with one or two male executors, and 2 cases in which two women were appointed as the only executors for an estate settlement.

Ownership by Women

Sole Ownership

Sole ownership by women was pioneered by widows who opted to remain active in a family fishing venture with which they were familiar and over which they exercised control. This pattern was followed early on by spinsters and then by married women, including a few even before the Married Women's Property Act was passed in Newfoundland in 1876. The first instance in the ship registers of a widow owning a ship was Ann McCarthy of Crocker's Cove, Conception Bay. She appears to have had the ship *Susan* built for her purposes by Richard Horwood of Moreton's Harbour in 1822. The ship was re-registered by Ann McCarthy in 1825. Her role in the registration may be ascertained from the wording of the original registration: "…having taken and subscribed the Oath required by this Act and having sworn that she is sole owner of the ship…*Susan*." The Act referred to is "An Act for the further Increase and Encouragement of Shipping and Navigation…passed in the 26th year of the Reign of King George the Third." The wording suggests that Ann McCarthy was physically present for the registration of her ship. The vessel was lost at Labrador, time unknown, and the ship register noted that the registry was closed on January 19, 1869. It is unlikely that the ship was involved in the Labrador fishery for all those years since registry of a ship was never closed until hard information was received, often many years after the loss or break-up.

The situation was similar with Catherine McGrath of Harbour Grace, widow, who as sole owner registered her ship *Success* in 1825. The ship had been built at Placentia as an open boat and "was raised on and decked in 1814." It was re-registered in 1830, which means that extensive rebuilding work had been done. James Fox was master of Catherine's ship.

Mary Woodley, widow (of Samuel?) of St. John's, was sole owner of the ship *Hope* when it was registered there in 1831. The ship had been built at Blandford, Nova Scotia, in 1825, and it was "rebuilt and enlarged" in 1831, the year that Mary registered it. From 1831 to 1841, when it was re-registered, the ship had 14 different masters. Did they resent working for a woman or was Mary a hard taskmaster? Mary also registered the ship *Two Brothers* in St. John's in 1831; it had been "built at Bareneed as an open boat in 1825 and never before registered." Samuel Woodley had been fishing out of St. John's since at least 1814, and Mary had opted to carry on the business alone after Samuel's death. According to manuscript records, Samuel probably died about 1830.

Nathaniel Woodley, Samuel's brother (?), died in 1833 and left his business to his wife Elizabeth, including "vessels, stages, wharves, etc." There is a manuscript record at the Newfoundland Provincial Archives of Elizabeth's business transactions from 1833 to 1857. Since Samuel had called one of his ships *Elizabeth* and Nathaniel had called one of his ships *Mary*, it is likely that this is a case of two brothers having married two sisters, or of an otherwise very close personal relationship. Elizabeth loaned money, borrowed money – including 546 pounds sterling and 100 pounds from Newman & Co. – rented her properties, bought land, paid insurance on two houses, bought sails, ropes, anchors, and tarpaulins for her ships, paid for "curing fish," and exported fish on three foreign-going vessels. Strangely, Elizabeth's name does not appear in the ship registers as a shipowner, but her husband Nathaniel had owned four ships: *Hero, Felicity, Brazilian Patriot*, and *Mary*, all registered between 1820 and 1831. Elizabeth must have held on to them in her deceased husband's name as a family company, though she sold the *Hero* in 1825 and the *Brazilian Patriot* in 1835.

Dinah Hollett, widow of Great Burin, was a shipowner sometime after 1819 according to family testimony, although I could find no confirmation of this in the ship registers. She was born in 1747, probably in England. John Hollett, her husband, was born in 1736 and was from Henstridge, England. Dinah and John were married in 1770, and their first child, Thomas, was born in 1771. They had five sons and two daughters. According to a letter from William Tulk, teacher at "the Bay of Burin school" in 1815 (a relative of John's), three of John and Dinah's grandchildren were attending Tulk's school. In fact, one of their sons had built the school. Tulk characterized Newfoundland as "this rocky wild and unchristian country," and described John as "the old gentleman" who is now sending Tulk's mother 30 shillings "in remembrance of Henstridge." One of the Hollett grandsons had been drowned while "coming home to Henstridge" the previous winter. John Hollett died in 1819, and a family member affirms about Dinah: "it would be around that time she may have taken over the responsibility of the shipping business, although her eldest son was 45 years of age when his father died." Actually Thomas was 48, but he may have been dealing with a strong-willed mother, accustomed to handling the business. Dinah's signature on her will of 1820 consists of an "X" but that does not mean that she did not have sufficient schooling to run a business. Many people who lacked formal education were reluctant to write their signature laboriously in the presence of an educated person making out the will and chose to defer to the "X" instead. Dinah Hollett died in 1826.

Mary Parsons, widow of Clown's Cove, registered the ship *Margaret & Sally* in 1834. It is obvious that she was present for the registration, for the record reads: "…having made and subscribed the declaration…and having declared that she is the sole owner of …" The ship had been built at West River, Prince Edward Island, and Samuel Parsons was the master. Mary had bought the ship on August 6, 1833, but it was not registered until 1834. See "Joint Owners," below, for the same Mary Parsons and for a different Mary Parsons.

Moving now to a later date, Charlotte Lee, widow of St. Jacques, Fortune Bay, whose husband had been Robert (?), the builder of her ship, owned the *Mary Jane* and registered the vessel in 1846, with Philip Kiddle as master. Catherine Walsh, widow of St. John's, registered the brigantine *William Stairs* in 1857. This ship had been built in Antigonish, Nova Scotia, the previous year. In its five years the vessel had only three masters, and it was lost at the seal fishery in 1862. Amelia Power, widow of Trinity, registered the 91-ton ship *Willie* in 1882. Charles Bugden was the builder and master. It was lost near the Horse Islands on November 12, 1889, probably on the way back from the Labrador fishery. Sarah Cluett, widow of Belleoram, whose husband had been William L. H. Cluett (?), the builder, registered the 95-ton vessel *Prima Donna* in 1884. This ship was broken up at Belleoram in 1915. Sarah must have been a woman of some influence since another vessel owned by Arthur and Levi Cluett was given her name in the early 1900s.

There is a most curious practice that prevailed at Burin in the period from the 1870s to the 1890s. Mrs. Eloise Morris, widow, ran a fishery and supply business during these years. She was likely related, possibly by marriage, to Patrick and Catherine Morris who owned ships and properties both at Burin and St. John's in the 1830s. She may have been the widow of Patrick, son of the original Patrick. She was the mother of three daughters: Lizzie, who married Capt. Hoberg, a Norwegian who had a small business in Burin and who was drowned on a voyage to Oporto; Kate, who died in Boston; and Ella, who was for many years the librarian at the House of Assembly in St. John's. Something of Eloise Morris' character may be gleaned from *The History of Burin* (1977): "Mrs. Morris, a very competent, well-educated lady of strikingpersonality and with a command of language peculiar to the old school, was rather imperial but not haughty in her manner. Being involved in a lawsuit which required her presence in the Supreme Court of Newfoundland in St. John's, she engaged as her lawyer the Late Lord Morris (then quite a young man). In her particular case he was obliged to show her all the old world courtesy such as engaging a cab, calling to take her to the court, offering her his arm, escorting her into the court after alighting first and assisting her down from the cab because she disliked being stared at by the 'rabble' around the courthouse door."

When Eloise Morris' customers in the Burin area went into debt, Mrs. Morris arranged a legal surrender of their property to her (ships, boats, shops, property, and so forth). Some say she "bullied" the men into signing over their property, but that may be too unkind a judgment. She became the legal owner of these properties, but there is no record in the ship registers of these vessels ever having been registered in her name. I have examined the records in the Burin Heritage Museum, and Fred Winsor has described Mrs. Morris' practices in his Ph.D. thesis, *The Newfoundland Bank Fishery: Government Policy and the Struggle to Improve Bank Fishing Crews' Working, Health and Safety Conditions, 1876-1920.* Her business involved fish, oil, fur, and lobsters, including the operation of a lobster factory. In the signed agreements, the fisherman was instructed to do everything for the benefit of "his said Master's interests" ("Master" meaning Mrs. Morris), and "according to the customs of the Fishery," i.e., the local customs of Burin. All agreements were signed and duly witnessed by her two office assistants, Michael Flynn and Hugh Reddy. Some of the agreements state that if they cannot pay their debt "we shall bring the said boat to her wharf if she requires it" as settlement of the debt. Customers were given from one to ten years to pay off debts; and in one case a man signed an agreement to pay an outstanding debt owed by his deceased father. At least five fishing boats are mentioned in the agreements: *Seaman's Pride, Mary Joe, Philomena*, and *Mystical Rose*; the fifth boat is not named. I was unable to find any of these in the ship registers, probably meaning that they were undecked boats and were not required to be registered. Fred Winsor claims that Mrs. Morris gained possession of 16 vessels in total. I expect that the original owners of boats and property continued to be supplied and financed by Mrs. Morris until they were able to pay off their debts. The fishermen who worked on Mrs. Morris' banking schooners (e.g. the *Laura May*) signed a different type of agreement, in which they were paid by getting "half their share of fish" but from this share they had to pay "their part of Bait, Ice, Freight, and Fishmaking according to the Custom of Burin." In fact, the schooner *Laura May* was signed over to Eloise Morris in 1890 for debts owed by the three male owners. Mrs. Morris was undoubtedly schooled by the merchant customs of her time to divert the profits to the business and not to the workers; she was a hard-nosed businessperson.

As a footnote to the preceding examples, it should be noted that in Vol. 22, Folio 80, at the Crown Lands Office, there is a transaction concerning the Morris family of Burin: "Eliza Morris of Burin widow in trust for the purposes declared in the last will of Mary Morris of Burin Spinster deceased and administratrix of the estate of the late Patrick Morris of Burin merchant deceased, her heirs and assigns…(land) known as Morris' Point, between Bull Cove and Ship Cove…7 acres and 8 perchas…$350…June 7, 1877." These were the incomplete notes I made in 2001, and I am not sure how they should be interpreted. Mary Morris was undoubtedly Patrick's sister. Were Eliza and Eloise the same person?

Mary Ann March of Old Perlican inherited two ships when her husband, Elias March, merchant, died without a will on January 18, 1908. The ships were the *Emily Bell* and the *New Daïsy*. Letters of Administration were granted to Mary Ann on May 1, 1908, and she sold both ships to William C. Job, merchant at St. John's, on November 18, 1908, having held them for little more than six months. Elias and his brother Eleazor jointly owned the business, but Eleazor, as captain of the ship *Dove*, was lost with all hands in 1871. The house flag was composed of two vertical red bars separated by a white bar. The business was bought and operated by one of March's clerks, George Howell. A nephew, Alfred March, son of Ebenezar, founded the March Shipping Line of Montreal. After World War II, this house flag was flown regularly by the Montreal firm.

The first spinster sole owner in the ship registers was Mary Foley Morris of St. John's. She may have been a sister of Patrick Morris whose widow Catherine was a shipowner in Burin and St. John's in the 1830s. Mary Foley Morris started her shipowning with a bang. She registered three ships in 1842: *John & Mary*, *Nancy*, and *Relief*. These had all been previously owned by Thomas Foley of Harbour Grace, undoubtedly a relative. She took mortgages on all three from merchants James Niner Wood and James Douglas. Margaret Blake, spinster of St. John's, was sole owner of the brigantine *Triton* when it was registered in 1845. She sold the ship in 1847. She appears to have inherited the ship *United Brothers* from her brother (?), Thomas Blake, and she had it rebuilt and re-registered in 1847. Patrick Sexton was the master.

Elizabeth Freake, spinster of Joe Batt's Arm, was the sister of Charles Freake of Birchy Bay, the builder and master of the 88-ton *Homeward* when it was registered in 1889. She owned all 64 shares so it is likely that she "staked" her brother to build the ship. Elizabeth Pennock, spinster of St. John's, in 1902 purchased the ship *Tasman* which had been built in Holland in 1893; it must have been a small luxury ship for the specifications mention crew space, captain's cabin, mate's berth, bosun's locker, storeroom, pantry and sail locker, and a house on deck. However, by the end of 1903 she had authorized Daniel Connally, master mariner, to sell the ship on her behalf. Gertrude Newman, spinster of Boyd's Cove, owned the 33-ton schooner *Margaret Newman* and registered it in 1910, the year that it was built by Joseph Newman, her brother (?). Joseph Butt, a fisherman at Joe Batt's Arm, held a mortgage of $800 for her at 5% per annum and it was payable by October 1, 1916.

Other spinster owners included: Evangeline Booth, who owned the Salvation Army ship *Bonavista* in 1899, originally named the *Salvationist*; Nina Osmond of Exploits, Burnt Islands, who bought the schooner *Phoebe M.* in 1936 and the *Lom* in 1938 and who carried on a business with her brother Otto for several years; Mary A. T. Roberts of Twillingate, who bought the schooner *Hattie A. Heckman* from Lucy Roberts, widow, in 1927; Mary Frances Bishop of Salmonier, who bought the *Mary and Bride Bishop* in 1927 (this vessel was condemned at Renews in 1952 as being unfit for further service); and Rosella A. Emberley of Creston, who owned the small schooner *Rose and Blanche*, which was built by her brother (?) William Emberley in 1928.

Sole ownership by married women started in 1885, nine years after the Married Women's Property Act was passed in Newfoundland.

The first married woman to own all 64 shares in a ship was Martha Mitchell of Grand Bank, wife of Charles Mitchell. Martha was a Hickman from Grand Bank. She registered the *Hattie E. Collins* that year. This ship had been built in 1866 at Booth Bay, Maine, U.S. The ship was active until 1888, and there is no information on her beyond that date. Mary Ann Quirk, married woman of Fortune Harbour, in 1893 owned the schooner *Notre Dame,* which had been built at Fortune Harbour by John Manuel of Kite Cove, almost surely for her specific use. Elizabeth Dean, married woman of St. John's, registered the *Switcher* in 1894; she sold it to Edward Maher in 1899. She also owned the *Isabella* for a time and sold it in 1901 to a master mariner in Nova Scotia. In 1895 Ellen Deen, married woman of Bonne Bay, registered the ship *Mary Ellen*, which had been built by Thomas Deen that same year. The small 12-ton ship was lost the next year "north-west of Fish Island Straits at Bellisle."

Margaret Stone of Rocky Brook, Smith Sound, Trinity Bay, registered the 37-ton *Cumberland* in 1901; the ship had been built by her husband, Emmanuel Stone, in 1898. On November 19, 1906, Margaret sold the vessel to Walter Baine Grieve, a St. John's merchant. Margaret is listed again later as a joint owner. Mrs. Mary Walsh of Harbour Breton bought the *Mary Sheehan* from French interests at St. Pierre-Miquelon in 1905, and it was lost in Fortune Bay just 20 days after she had purchased the ship. Bridget Mahoney, married woman of Bay Roberts, registered the vessel *Birch Hill* in 1911; it was "totally lost" or "cast away" on October 14, 1916, on the way back from Labrador. For more information on this case see the later section on "Ships, Irregularities and Even Hints of Scandal." Grace Petipas of Summerside, Bay of Islands, in 1911 bought the *Henry M. Stanley*, which had been built in Essex, Massachusetts, in 1899 and rebuilt in Bay of Islands in 1911. It was a Labrador-going ship, and she sold it on July 4, 1913, to merchants in Halifax.

As the twentieth century approached, shipowning by women seemed to have become a more natural activity. Bride (or Bridget) Goff, married woman of St. Joseph's, bought two small ships in

1898, the *Hero* and the *Florence Silver*. Rachel Johnson, married woman of Little Catalina, became owner of the *Rego,* which her husband William built in 1899. She sold it the next year to a trio of merchants in Catalina but bought it back from them in 1905. Mary Keech, married woman of Epworth, bought the 55-ton *Jessie M* in 1904, held onto it for six years and sold it to a merchant in 1910. The Keech family later moved to western Canada.

In 1902 Anastassia McDonald of Salmonier owned two ships: the *Bonny Lass* and the *Senator*. John McDonald had bought the first vessel in 1898 and transferred title by bill of sale to Anastassia in 1902. He did the same thing with the second vessel, which he bought in November 1902 and sold to Anastassia one month later. This may well have been a husband-to-wife transfer, but there seems to be some modeling going on here, which is evident in 1909 when Anastassia sold the *Bonny Lass* to Ellen McDonald, married woman of Salmonier. Ellen then bought the *Anita C* from Michael F. McDonald in 1922. Anastassia McDonald died in September 1917 at Newbridge, Salmonier, leaving a husband, three daughters and two sons.

Theresa Stone, married woman of Catalina, bought the 20-ton *Theresa S.* from its builder William Johnson in 1906, naming the ship after herself and holding on to it for eight years. John House of Gooseberry Island built the *Lizzie H.* in 1902, sold it to St. John's merchant William C. Job in 1903, bought it back from Job in 1905, sold it to Emma House (very probably his wife) by bill of sale in 1906, and she sold it to George Knowling, St. John's merchant, in 1910.

Jane Lawrence, widow of Alexander Lawrence of St. John's, was the owner of four ships. Alexander died in 1906 and James Lawrence was appointed executor of the will. In 1908 he sold the 43-ton *Glencoe* to Jane, which she used for nine years, the *Jim L.,* on which she took a $2,000 mortgage in 1911 and which she sold in 1913, and the *Emily*. In 1918 James Lawrence sold the 63-ton *Lewisport* to Jane, and she sold it the same day to Joshua Winsor of Wesleyville. A quick sale indeed!

Ellen M. Shea, married woman of St. John's, bought the 142-ton *Artist* from Patrick J. Shea two days after he bought it in August 1916. The vessel was in a collision with the S.S. *Espanolets* of Malaga in the Spanish Mediterranean on February 20, 1917 or 1927; the date is indistinct in the records.

Catherine Greene, widow of Cape Broyle, bought the 44-ton ship *Effie M.* in 1921 under Mortgage B (the second mortgage), which had been taken out by the previous owner, Jesse Winsor of Wesleyville. Catherine Fewer was born at Chapel's Cove, Harbour Main, in 1855. In 1880 she became the second wife of Constable Thomas Greene (his first wife had died in childbirth). Catherine and Thomas had thirteen children; her eldest son Walter was killed in 1917 during World War I. Thomas died in 1914 while stationed on Bell Island, and Catherine returned to Cape Broyle. She may have bought the ship for her son Greg, who ran a general store in Cape Broyle. According to family testimony, although she had little education, she carried herself like a queen and acted the part of a refined lady. She went blind in her late years and died in October 1946 at age 90.

Eliza Reid was born at North Harbour in 1885 and in 1900 married Josiah Pafford, a planter/trader in that place. They had twelve children. Eliza had a Grade 8 education and was a co-partner in the business with her husband and was the book-keeper. Josiah died in 1924 with typhoid fever, as did five of their children, and in 1928 Eliza married George H. Eddy, who operated a general store, packed herring for export, and was in the salt-cod business. They continued to use the small schooner *L. G. P.*, which Josiah had left to Eliza and which was named for their two daughters, Lillian and Gertrude Pafford. Not until 1931, seven years after Josiah's death, did Eliza seek Letters of Administration to regularize ownership of the schooner. She obviously did this in order to sell the ship. Letters were granted to her on May 19, 1931, and she sold the ship nine days later to William Farrell, master mariner of St. John's.

The length of time that women held on to ships is of special interest. Here is a sampling from the records. Ellen Hann of Petit Forte bought the *Christie M.* from the Marystown Trading Co. Ltd. in 1924 and held on to the ship for almost 11 years. Ellen Whelan was born in 1879 at St. Brendan's, Bonavista Bay, graduated with teacher training from St. Bride's College, Littledale, and went to Petit Forte as a teacher, later marrying Michael Hann. They had nine children. Ellen owned the ship *Christie M.* from 1924 to 1930. She was involved in the family business as secretary and book-keeper. She died in 1955. Irene Jane Blackmore, married woman of Port Union, bought the *W. J. Phillips* in July 1926 and sold it in July 1937, exactly 11 years. Elsie T. Roberts, married woman of Wesleyville, bought the *Lillian and Jackie* in March 1928 and sold it in June 1939, that is, after 11 years and 3 months. Elizabeth Jane House, married woman of Catalina, inherited the *Lizzie Guy* from Mark Guy (probably her brother) and held on to it for almost 12 years.

Eliza Matilda Pardy, whose husband was Capt. Thomas W. Pardy of Harbour Breton, bought the 23-ton *Maraville* in January 1930 and sold it in October 1943, having used it for 13 years and 9 months. Eliza Sibley was born in 1881 at Harbour Breton. She and Capt. Tom were married in 1904 and they had two children. Eliza died in 1967 at age 86. Annie (Newman) Hannam, whose husband was Johnnie of Rose Blanche, bought the *J. A. Hannam* in October 1931 and sold it in September 1946, that is, 15 years less 1 month.

Sarah Blaikie Ayre inherited the 32-ton *Cactus* when her husband John Bray Ayre died in April 1915, and she sold it in December 1930, that is, 15 years and 8 months. Lydia Hollett of Burin, bought the *James Young* from Hollett's Fisheries Ltd. of St. John's in 1931, had mortgages on the ship for 13 years, and sold the ship after more than 17 years of use. Lydia Foote was born at Step-a-Side, Burin, in 1872 and married William Henry Hollett in 1897; they bought a business at Burin North in 1914, had three children, and Lydia died in 1957. Jessie Bragg, widow of Port Union, held on to the ship *Marie Yvonne* for 21 years (1943 to 1964), and

Maud Billard, with her husband Simeon, held on to the two ships *Dolores Kaye* and *Austin & Dianne* for a total of 36 years.

There are other cases which are of interest. Harriet (Edwards) Billard of Rose Blanche, whose husband was Samuel, a fisherman, owned the 25-ton *Ethel & Albert* from 1932 to 1939. She was an aunt of Maud Billard of Harbour Le Cou, also a shipowner, so again we see an example of modeling shipowning from one generation to another. Annie Hickey of Harbour Breton, whose husband was Augustus Hickey, bought the 12-ton ship *Eleanor & Barbara* from the Stone Brothers of West Point, La Poile, on August 16, 1951. Augustus was a trader in the Bay de Loo, Rencontre East area. Annie ran the store and was in charge of the business in Harbour Breton. Annie's birth name was Sarah Ann Bullen and she was born in 1912 at Lamaline. She and Augustus had eight children; they later moved to Wood's Island, and she died about 1992.

Jane Bartlett, married woman of Rattling Brook, owned four ships over a period of nine years in the 1930s and 1940s: the *Arizona* from 1932-34, the *Radio City* from 1935-36, the *Audrey Bartlett* from 1936-37, and the *Tempest* from 1940-42. Jane's husband, William Francis Bartlett, was a merchant and promoter of shipbuilding in the Green Bay area; all four ships were built under his auspices. Julia Gilbert, married woman of Haystack, Placentia Bay, bought the 33-ton *Bride and Annie Dunphy* in August 1937, installed an engine in it in 1939, and sold it in April 1946. Gertrude (Yarn) Evans, whose husband Tom managed the H. E. Petite business firm in Mose Ambrose for several years, bought the 34-ton *Jessie Beatrice* from Ada Annie Petite in 1937, installed an auxiliary motor screw in the ship in 1938 and sold it in 1942 to Jerry Petite & Sons Ltd. of English Harbour West. Mary Vallis, married woman of Coombe's Cove, Fortune Bay, bought the *R. J. Devereux* in 1941 and changed the name to *Mary King* (her birth name?). The vessel was broken up at Harbour Breton in 1952, so it can be assumed that it was in use until that time.

There were two women on the south coast of Newfoundland who are deserving of special mention. The first is Ada Annie Petite of Mose Ambrose. Ada Annie Hill was born at Sandy Point, St. George's Bay, Port au Port Peninsula, in 1887 and was teaching at Spaniard's Bay when Harry Edwin Petite, a young widower from Mose Ambrose, came to town on business and just happened to meet the local school teacher. They were married in 1912, and their son Gordon was born in 1920. At the point where the family business, H. E. Petite Ltd., had reached a level of prosperity where Harry could spend most of his time at home, he was drowned in 1925 when an open freighter taking supplies from English Harbour West to Mose Ambrose swamped in heavy seas and high winds with all hands lost. Ada Annie was described as "a well-read woman of strong character." Using all her personal resources she kept the family business going, including the bank fishery, with the help of many hired servants and ships' crews. She supervised every aspect of the operation and set an example of hard work and unique resourcefulness. In 1928 she bought the Gloucester-built ship *Charlie and Eric*; it was lost off Cape Harrigan, Labrador, in 1933. In 1934 she bought the 129-ton *Rex Perry*, one of the first power vessels on the coast; it was lost in 1952 off Miquelon Head. And in 1937 she bought the 34-ton locally built ship, the *Jessie Beatrice*; she sold this vessel to Gertrude Evans of Mose Ambrose in 1938. She also owned the schooner *Uncle Bob*. Ada Annie built up a very prosperous business and kept it going until her son Gordon was old enough to take it over, and in 1953 he bought out Jerry Petite & Sons Ltd. of English Harbour West and operated both firms. In that same year, 1953, Ada Annie went to England for the Queen's Coronation. The Mose Ambrose directory of 1936 described Ada Annie Petite as "general merchant, vessel owner, importer of British, American and Canadian goods, exporter of codfish, cod oil, lobster, salmon: Dealer in dry goods, groceries, provisions, hardware and fishery supplies." In addition to the store at Mose Ambrose, she also operated branch stores at Red Cove, Boxey and Coomb's Cove. In 1954 Ada Annie moved to Toronto to live with her daughter Ada and died there in 1964. The firm at English Harbour West is operated today by Debbie Petite, Ada Annie's granddaughter.

Deserving a category of her own is shipowner Marie Smart Penny. Marie Smart was born at Little Bay, Notre Dame Bay, and graduated from the Channel Anglican Academy, later taking private tutoring at Carbonear; she subsequently graduated from St. Bride's College, Littledale. She married George J. Penny, manager of John Penny & Sons Ltd. of Ramea, in 1927. When her husband, by then Senator Penny, died in 1949, Marie became the president of the Penny Company. She was at one time the owner of ten vessels and was familiarly known as "Queen of the Fishing Fleet." She had more than 200 workers in her employ and knew every one of them by name. In 1970 her company shipped four million pounds of processed fish to the United States. Marie Penny's daughter Margaret, a graduate of Mount Saint Vincent School of Commerce, and her son-in-law Spencer Lake served with Marie and four other members of the Board of Directors to continue a most successful business along the south coast for many years. In 1959 she was one of only 35 Canadians selected to dine with Queen Elizabeth and Prince Philip in Ottawa. At her home in Ramea, known as Four Winds, she once entertained Governor General Vincent Massey, as well as many other dignitaries and ordinary visitors. In 1967, in recognition of her abilities and business acumen, she was elected the first woman president of the Fisheries Council of Canada, a 400-member organization. Most of the Penny ships were registered under the company name, and Marie Penny is mentioned only once in the ship registers when in 1954 she and Spencer Lake bought the *Blue Surf*, a 178-ton ship from Job Bros. and sold it later that year to Burgeo Fish Industries Ltd. In the records both Marie and Spencer Lake are titled as "Business Managers." Marie Penny died in 1970.

Ownership by married women prevailed over ownership by widows and spinsters, particularly after about 1900, and it soon became the dominant form of sole ownership by women.

For more women Sole Owners see Appendix A.

Joint Ownership

The second most favoured method of owning ships was joint ownership in which a woman was one owner of two or one owner among several: a widow owning with her deceased husband's partners (usually his brothers); a woman owning with her husband; a single woman owning with family members; a woman owning with several other males; a woman owning with one or more women; and women purchasing one or more of the 64 shares in a ship.

Margaret McCarthy, widow of Carbonear, was a joint owner with John McCarthy, master of the ship *George* when the vessel was "raised on and enlarged" in 1823. Eric Gosse of Spaniard's Bay stated in a 1995 letter to me that Margaret and John were joint owners of this vessel as early as 1804, which would have been before ship registration was regularized in Newfoundland. There are other references to registrations of the *George* in 1812 and in 1826. Interestingly, at the top of the original registration is written: "D. Buchan, Capt. of Ship *Grasshopper* & senior naval officer." This was David Buchan of Beothuk fame, but it seems from the records that he did not come into possession of the *Grasshopper* until 1819, so another registration may have occurred that year. The role of the woman in the act of registration may be ascertained from the wording of the original agreement. Of John McCarthy it says: "...and having sworn that he together with Margaret McCarthy also of Carbonear aforesaid widow who is not within 20 miles of this place and who has not absented herself to avoid taking the Oath required..." Presumably the registration took place at St. John's, or possibly at Harbour Grace. Compare the previous wording for Ann McCarthy, the first of the widow sole owners, which is even more specific.

Mary Parsons, widow of Clown's Cove, was listed earlier as a sole owner of the *Margaret & Sally* in 1834. But 16 years before that, in 1818, she was a joint owner of the 35-ton schooner *Venus* with Edward and John Parsons, planters and builders of the vessel. It is just possible that this may have been another Mary Parsons; the 16-year gap seems curious. Edward Parsons was the master of this vessel. He and John were probably her brothers-in-law.

Another Mary Parsons of Clown's Cove made an appearance in 1848, but this one was the wife of Alfred Parsons. Alfred was a master mariner and probably was away when the two merchants who held his mortgage on the 38-ton schooner *Zephyr*, Charles F. Bennett and Thomas Bennett, transferred that mortgage to two other merchants, William H. Taylor and John Bemister. The two new merchants are noted to be "Trustees to Mary Parsons Clown's Cove wife of Alf.d Parsons of Newfoundland, Master Mariner." The note was not added to the record until May 20, 1854. Alfred must have been away on a prolonged sea voyage. On his return Alfred lengthened and enlarged the ship at New Perlican in May 1855 and registered it *de novo* in 1856. It is possible that Mary may have been continuing a local fishing venture in her husband's absence, under the watchful eyes of the merchants, or she may have fallen behind in her obligations to them.

Olive Hederson, widow of James of Brigus, was a joint owner with Thomas Hederson of the ship *Twin Brothers* when it was registered in 1823. The same note about Capt. David Buchan appears at the top of this registration form. Elizabeth Pitts, widow, was a one-third owner with James and William Pitts of Belle Isle, Conception Bay, in 1825 when they registered the *Belinda*. That means each of them owned 21 1/3. Elizabeth was probably the widow of John Pitts. The vessel was lost in 1836.

Mary Taylor, widow of Carbonear, owned 32 shares of the schooner *Industry* when it was registered in 1826. Thomas Chancey, merchant of Carbonear, was the master and owned the other 32 shares. In that same year Elizabeth Taylor, widow of Carbonear, was a joint owner of the ship *Active* along with William Powell, the master, merchant Robert Pack of Carbonear, and John Gosse of Poole, and William Fryer of Wimbourne, England. Taylor and Powell owned 21 shares each, and Pack, Gosse and Fryer owned 22 shares among them. Elizabeth Taylor was the "Administratrix to the Estate and Effects" of the late Nicholas Powell, so she probably was his sister. The vessel was broken up in 1829.

The merchant Simon Levi and Ann Toque, widow of Philip Toque of Carbonear, were joint owners of the ship *Elizabeth* when it was registered in 1831 (they held 32 shares each and were designated "co-partners in trade"). Levi and Philip Toque had been partners when they registered the 105-ton schooner *Alpha* in 1828. Philip must have died before 1831 since Levi and Ann Toque were co-partners in December 1835 when they sold the ship to St. John's merchants. Levi was a Jew and in the election year riots of 1836 his house was attacked twice under the accusation that he was an informer. When Levi died in 1837, the trustees of his insolvent estate transferred all his shares to John Rorke, merchant of Adams Cove, and Ann Toque continued as Rorke's "co-partner in trade" when they re-registered the ship in 1839. Ann was the daughter of Robert Howell of Carbonear and is listed with her three brothers in the *Conception Bay Plantation Book* of 1805 as inheritors of their father's property there in 1800. Of additional interest is the fact that Ann and Philip Tocque's son, also named Philip, became a literary figure of some note. While still a young man in Carbonear, he and John Elson and others founded a literary circle. In 1849 he moved to Boston and became involved in several liberal causes, including the rights of women. Obviously, his mother's proactive approach had made a strong impression on him. By 1852 young Philip had moved to Nova Scotia, and in the 1870s he was living in Toronto.

Catherine Morris, widow of St. John's and Burin, was joint owner with Joseph Ryan, merchant of St. John's, of three ships, *Shamrock, Saint Patrick,* and *Calypso*, when these were registered in 1831, 1832 and 1833. They are designated as "co-partners in trade." The first vessel was lost in 1834; in 1835 Catherine sold her shares in the second vessel; the third vessel was still operating in 1837. Catherine was the widow of Geoffrey Morris, and the firm of Morris & Ryan of Poole and Liverpool had been operating in Burin from at least 1812.

Charles Simms of St. John's, barrister, and Elizabeth Henderson, widow of Harbour Grace, registered the British-built brig-rigged ship

Elizabeth in 1837. Each of them held 32 shares, and they re-registered the ship in 1844, obviously as the result of modifications and some rebuilding. Ann Elizabeth Henderson died in 1858 at the age of 68; she had been the wife of Jordan Henderson, merchant at Harbour Grace. Charles Simms was her son-in-law. Richard Brown, planter, and Catherine Brown, widow of Burin, jointly owned the ship *Catherine* when it was registered in 1844. Richard had been the builder. He had to take a mortgage on his 48 shares from a St. John's merchant, but Catherine did not take a mortgage on her 16 shares.

Twenty-three years before the Married Women's Property Act was proclaimed in 1876, Julia Kelligrew was a married woman shipowner in partnership with her husband, William Kelligrew, and John Goodridge, both merchants at St. John's. In 1853 they registered their ship *Ann*, with William and John holding 48 shares together as "co-partners in trade," and Julia holding 11 shares each for her under-age children John William and William George. This was obviously Julia's way of providing for her children's future. Unfortunately, the vessel was stranded and declared lost near North Sydney in 1854 and was sold by public auction. The Kelligrews were in business at St. John's until at least 1875.

Ann Giles, widow of Carbonear, was a partner with William Giles, John Giles and Richard Giles when they registered the ship *Nelson* in 1853. Each owned 16 shares. The ship was broken up at Carbonear in 1870. In 1855 Robert Pack of Carbonear registered the ship *Christina* and his partners were "...John Gosse of the Town of Poole in the County of Dorset, merchant, and Mary Rogers Fryer, widow, and representative of the late John Fryer of Wimborne in the County of Dorset, merchant..." This vessel was lost in 1862.

Catherine Stoneman, widow of St. John's, was a partner with George Stoneman of Trinity when they registered the ship *Henry Thomas* in 1858. Catherine's husband William had died at Trinity in 1856, so Catherine had obviously moved to St. John's in the interim. Their son, William R. Stoneman, became a writer and was the author of a book, *The Penitent's Tale: Or, The Backslider's*

Experience, published by the *Morning Post* in St. John's in 1848. The Stonemans had been living at Trinity since at least 1828, and William had owned six ships during that period, including one named *Catherine*. The *Henry Thomas* was lost in 1866 on a voyage from North Sydney to Newfoundland.

Catherine Cummins, widow of St. John's, was a joint owner with Peter Cummins when they registered the ship *Kate Cummins* (named after Catherine) in 1862. Catherine's husband Maurice was a prosperous merchant-planter of St. John's and had died in 1853 at age 66. Their daughter Catherine had married J. B. Sawyer, Esq., in 1853, and a second daughter, Lizzie Agnes, married Alexander Murray, Esq., of the Geological Survey in 1868. The *Kate Cummins* was lost in December 1870 at Cow Bay, Conception Bay.

In 1844, 22 years before the Married Women's Property Act became law, Caroline Winsor, married woman of Aquaforte, owned shares in the ship *Pelter* with her sister-in-law, Jane Winsor, and three Winsor males. In 1866, Elizabeth Pond of Greenspond was a partner with her husband John and his brother Andrew when they registered the *Edward & Lydia*. They held the 64 shares jointly. She had been Elizabeth Wells and had married John on November 4, 1835; they had no children, so she may well have been significantly involved in the fishing venture.

Ellen Roach of Branch, St. Mary's Bay, was designated as a "Trader" in 1887 when she and her two partners registered the ship *Fleetwing* at St. John's. Her partners were Henry Roach, her husband and builder and master of the vessel, and his brother William Roach. They were designated as fishermen. Ellen's birth name was Cantwell and she was born at Torbay in 1852. Perhaps she went to Branch as a school teacher. She married Henry Roach in 1877, and they had nine children. The term trader usually refers to coastal trading of goods and picking up dried cod on these visits to communities. In this case it may refer to the fact that Ellen was the business head of a general store and accompanied her male partners in some of these trading ventures. This was common along the south coast. Ellen died in 1931.

At a somewhat later date, the Tessier women of St. John's became involved with ships. Julia Tessier was a married woman in 1893 when she and Charles Nicholls, a master mariner of St. John's, registered the 207-ton British-built ship *Petunia*. This was a three-masted brigantine-rigged ship designed for foreign-going ventures. Julia held 48 shares; Charles Nicholls, master, held 16 shares. Julia died without a will late in that same year and Letters of Administration were granted by the Supreme Court the next year to James C. Tessier, and he sold the ship to George James Tessier, accountant. Prior to Julia's involvement with the *Petunia*, Ann Catherine Weston Tessier, widow, had owned the ship with Peter Germon Tessier, Lewis R. Tessier, Charles W. Tessier and Weston G. Tessier. They had sold it to Julia and Charles Nicholls. Ann Tessier had owned shares in the *Native* in 1886, in the *Lilac* in 1891, in the *Mistletoe* in 1893, and in the *Brisk* in 1900. Anne C. M. Tessier, spinster and daughter of Peter Germon Tessier, owned 3 shares in the *Mistletoe* in 1893 and bought a further 10 shares in 1895; she owned 9 shares in the *Silver Sea* in 1895 and bought a further 11 shares in the same ship in 1902 and held 7 shares in the *Gladiola* in 1898. Bertha Mary Tessier, married woman, bought 4 shares in the *Silver Sea* in 1893 and 1 additional share in 1895 and 33 shares in 1902. She bought 3 shares in the *Mistletoe* in 1896 from Anne C. M. Tessier. She held 3 shares in the *Gladiola* in 1898. All four Tessier women were speculators in the ship business.

A pattern of speculation emerges even from the earliest records. The Winsor women of Aquaforte both owned shares in the ship *Pelter* as early as 1844. Charlotte Colelough, spinster of St. John's, bought shares along with 15 other people in the ship *Sybil* in 1852. Charlotte is also noted to be a "co-partner in trade." Ruth Orr Church, married woman of St. John's, purchased 1 share in the *Frances E. Moulton* in 1923, and Meta Boals Orr purchased 5 shares from her husband, John B. Orr, in the same ship in 1926, 3 shares in the *Ena A. Moulton*, 3 shares in the *Catherine M. Moulton*, 1 share in the *Enid E. Legge*, and 1 share in the *Jean Wakely*, all in the same year: 1926. This pattern continued. Eva McIsaac, married woman of Searston, Grand River, bought 1 share in the *Jean Wakely* in 1920 and was still holding it in 1925. Elizabeth Findlater, married woman of St. John's, bought 1 share in each of the *Ena A. Moulton*, the *Enid A. Legge*, and the *Jean Wakely* in 1921.

Mary A. Rose, merchant of Harbour Breton (a later hand crossed out this occupation and wrote in "married woman"), bought 1 share in the *Gordon E. Moulton* in May 1921 and sold it in August 1922 to John T. Moulton. Mary Stewart was from Burgeo; she married John Rose of Little Bay and they settled in Harbour Breton; Mary was always referred to formally in the community as "Mrs. John Rose"; they had no children. They were the owners of the famous Sunny Cottage (the community museum today), and together they ran a business which consisted of buying and shipping fish, selling groceries and antiques, and coasting – carrying lumber and picking up salmon on the Labrador and taking it to North Sydney. Mary assisted with the business.

Then there were the Grand Bank women who are worthy of note as speculators and joint owners. There was no other place in Newfoundland that saw such a concentration of women buying shares in ships. Women buying shares occurred particularly after 1920 when the large three-masted foreign-going ships were being built in Grand Bank and the surrounding areas. It is apparent that the women of Grand Bank wanted in on the action.

Sarah W. Clarke bought 3 shares in the *Max Horton*, William Forsey's ship, in 1920. Her daughter, Georgina T. Clarke, stenographer, also bought 3 shares. Incidentally, Sarah's husband, Lionel, merchant, also bought 3 shares, and her son Roy, clerk, bought 2 shares. Sarah was a Tucker, originally from Carbonear. She also held 5 shares in the *General Trenchard* in 1923. Georgina later married an Elliott and settled in Corner Brook. Charlotte Pratt Harris bought 5 shares in her husband George Chesley Harris' ship, the *General Woods* in 1921 and another 2 shares in the same ship in 1926. She was still holding these shares in 1938. Charlotte was the sister of E. J. Pratt, famous Newfoundland poet, and the daughter of Rev. John Pratt, Methodist minister, who died during his tenure at Grand Bank. Charlotte was a trained singer, had no children, and died suddenly at age 70 on September 8,

1956. Sophie Knight, spinster, bought 2 shares in the *General Woods* in 1922; she died in 1926. Her brother Archibald was one of the executors of her will.

Emily Dunford, married woman, bought 1 share in the *General Woods* in 1922 (which she was still holding in 1938) and 1 share in the *General Plumer* in 1927. Her husband was George, a salesman. Emily was the daughter of Samuel and Mary Harris and sister of Eleanor Carr and Mary Forsey Harris. She attended Mount Allison University, had two children, and died on October 22, 1948. Mary E. Kelland, widow of Lamaline, (we will consider her here as a Grand Bank woman) purchased 1 share in the *General Smuts* and was still holding it in 1925, the last date of record. Priscilla Noseworthy of Fortune (a Grand Bank neighbour) bought 21 shares in the *Alhambia* in 1930 and sold them two years later. She was obviously a speculator. Mary Forsey Harris, spinster and daughter of Samuel and Mary (Forsey) Harris bought 5 shares in the *General Plumer* after her father, the owner, died in 1926; in 1927 she bought 5 shares in the *General Trenchard*, and she held 1 share in the *General Woods* from 1927 until at least 1938. She also attended Mount Allison University. She died in Toronto in the 1970s.

Harriet Marion Harris, widow of Grand Bank, purchased 5 shares in the *General Gough* in 1927; she was still holding these in 1931. She bought 2 shares in the *General Woods* in 1927 and was still holding these in 1938. Harriet was the daughter of George Harding of Grand Bank and she became the second wife of Samuel Harris after his first wife, Mary, died in 1913. Harriet Marion died on February 25, 1951.

Eleanor Hickman Harris Carr, married woman, bought 1 share in the *General Woods* in 1927 and was still holding it in 1938, and bought 3 shares in the *General Plumer* in 1927. Eleanor Carr graduated from Mount Allison University with a B.A. and then earned her R.N. at the Massachusetts General Hospital. Percival Carr had come to Grand Bank as treasurer and business manager of Samuel and George Harris' business firm. Eleanor and Percy were married in 1920. They had six children and owned the famous Clawbonnie Estate, midway between Grand Bank and Fortune. When her husband died in 1943 Eleanor moved to Toronto and later worked as a nursing supervisor in an Ottawa hospital. She died in Toronto in 1975.

Jane Forsey, married woman, bought 2 shares in the *General Woods* in 1937 and was still holding these in 1938. Her husband was Capt. Steve Will Forsey. Jane was a Banfield and was born at Garnish. Mary Durey of Marystown inherited 1 share in the *General Woods* when her husband Patrick died in 1937; she was still holding this share in 1938.

There were other Grand Bank women who held shares in ships. Sarah Patten was the daughter of John B. Patten and Elizabeth (Hickman) Patten and was born in 1895. After she matriculated in Grand Bank she attended Shaw's Business School in Toronto. She returned to Grand Bank to take charge of the office work for the Samuel Harris firm. She then became the accountant at her father's business, J. B. Patten & Sons. In 1927 she married George Foote and they had five children. When her father died in 1927, she and her brother were appointed executors. They jointly held the 32 shares which they inherited in the *Lillian M. Richard* and sold them after three years. In 1948 Sarah bought 9 shares in the *Arawana* and re-sold these after four years. Also in 1948 she bought 16 shares in the *A. & R. Martin*; this vessel was lost off Lamaline in 1952. She moved to London, Ontario, with her husband in 1981 and died there in 1982 at the age of 87.

Esther Linda Foote, married woman, bought 10 shares in the *Anton C. Santos* in 1927 and was still holding these in 1939. Esther was the daughter of Simeon and Grace Tibbo; her husband was Ambrose. They had five children and she died in 1968. Emma Louise Foote was the widow of John B. Foote when she bought 9 shares in the *Anton C. Santos* in 1927. She died without a will in 1935 and Letters of Administration were granted to her brother-in-law, Ambrose Foote, and he sold the shares. Emma was the sister of Charles Forward of the firm of Forward & Tibbo. She and her husband had five children and she died in 1935.

For more women Joint Owners see Appendix A.

There were several varieties of joint ownership and it will be useful to look at some of these.

Owning Ships Jointly with Husbands

The Newfoundland ship registers do not make clear the husband-wife relationship when these are joint owners. In many cases, confirmation had to be sought elsewhere; and in several cases it is a best guess. When these cases are a duplication of previous and subsequent references, a concerted effort has been made to give only minimal information.

Caroline Winsor of Aquaforte was a partner with Jane Winsor and four Winsor males in the ship *Pelter* in 1844. It is known that Caroline's husband was Dan Henry Winsor and one of the male owners is Henry Winsor, a master mariner. It is assumed that this is Caroline's husband. On the other hand, it is known that Julia Kelligrew and her husband William, along with John Goodridge, were joint owners of the schooner *Ann* in 1853. The same is true of Elizabeth and John Pond of Greenspond in 1866, as we saw earlier, in their ship *Edward & Lydia*, but the marriage record had to be ascertained from another source.

It is also assumed that Ann J. Tuck of Fortune is the wife of John Tuck, the builder and captain of the *George A. Tuck* in 1886. Ellen Roach, trader of Branch, St. Mary's Bay, was a partner with William and Henry Roach in the ship *Fleetwing* in 1887; Ellen was the wife of Henry, the builder and captain. So also with Emma Wheeler and Symeon Wheeler of Frenchman's Cove, Bay of Islands, in 1896, with their vessel *Spring Bird*. Martha Benson, married woman of

Little Bay Islands, owned the *St. Elmo* with Reuben, Arthur, and Jonathan J. Benson when it was first registered in 1898 and re-registered in 1912. One of these men was almost certainly her husband, probably Jonathan, since Martha's name appears as the second owner, immediately after his in the register. Each of the four held 16 shares. They sold the ship on June 12, 1914.

Margaret Kennedy, married woman of Carbonear, and William John Kennedy, planter, bought the 106-ton *Pioneer* from William A. Munn of St. John's in 1900; it was abandoned at sea on December 18, 1902, off the Labrador coast. That same year Margaret sold her ship *Pride of the North* to Dugald Munn. By March 1913 Margaret was a widow and she bought the *Winnie F. Tuck* from Patrick J. Shea. This vessel was "totally lost" at Emily Harbour, Labrador, on October 1, 1919. It seems safe to assume that William John had been Margaret's husband. It should be noted that even after she was widowed she continued to be involved in the Labrador fishery.

Margaret Stone of Rocky Brook, Trinity Bay, who has already been discussed as a sole owner, must have been an entrepreneur, because in 1906 she bought 6 shares in the *H. W. Stone* and she also owned 11 shares in the *Maggie Stone*, and her husband Emmanuel owned the other 53. Mary Margaret Cooper was born at Bluff Head Cove, Trinity Bay, in 1861. She was the grandmother of Capt. Harry Stone, who described her in 1995 in the following way: "She was all go, she was very involved in everything, a very determined woman; she would take a horse into the woods for a load of wood; in the spring of each year she would walk from Rocky Brook to St. John's; it would take a week; she had little education but she could read." The walk from Rocky Brook to St. John's must have been in the very early years, because the *Free Press* for 1912 to 1914 indicates that Margaret and her husband were making regular spring and fall trips to St. John's in their schooners to purchase summer and winter supplies, sending their loaded schooners home, and taking a leisurely trip home themselves by the express train. In March 1914 Emmanuel brought Margaret to St. John's for surgery at the General Hospital. She died in 1918 at age 57. Curiously both she and her husband died without wills.

There is a long hiatus between the Stones and other married couples becoming joint owners on the south and south-west coasts. There it became commonplace for husband and wife to be joint owners. Joint ownership by husband and wife is particularly focused in the Rose Blanche and Harbour Le Cou areas. In 1944 Gordon Lewis Skinner and Pearl Melita Skinner of Rose Blanche were joint owners of the *Doris A. Neal*, a fishing schooner. Gabriel Billard and Bessie Billard of Rose Blanche bought the *Calvin Anne* in 1943 and re-sold it in 1948 to George Robert Durnford and May Durnford of Francois. Though her husband died in 1949, May held on to the ship until 1960 when she sold it. Edwin H. Hardy and Clara Theresa Hardy of Rose Blanche bought the *Emma H.* in 1945. Clara had been designated as "housewife" in the original registry, but this was crossed out by a later hand and "married woman" was written in. Martin Collier and Mary Elizabeth Collier of Rose Blanche bought the *Chesley Russell* in 1943 and sold it in 1956, having held it for a total of 13 years. John William Wells and Alice Wells of Rose Blanche bought the *Melita Wells* in 1944 and sold it in 1955. Already noted in Chapter 1 is the case of Henry and Myrtle Hatcher of Rose Blanche, and there were at least six other married couples in this small town who owned fishing ships jointly and used them in the fishery for periods up to 12 years. For these cases see Appendix A.

In nearby Harbour Le Cou the situation was similar though the numbers were not as great. There were only five women ship-owners in the small fishing village between 1926 and 1983, but the pattern of married couples owning a ship jointly is evident here also. Augustus Buckland and Phoebe Buckland bought the *Dolores Kaye* in 1943 and sold it to another married couple in 1947, Simeon Matthew Billard and Maud Billard, also of Harbour Le Cou. In 1946 the Bucklands bought the *Carolyn Marie* and sold it in 1959. Phoebe was a Herridge, and she and Gus had seven children. She arranged for the sharemen who worked with Gus to live in the house with the family during the fishing season, fed them, did their washing, and worked on the flakes making fish. In 1963 Simeon Billard registered the *Austin & Dianne*, a longliner of which he had been sole owner since 1960, jointly in his and Maud's names, and they used the vessel until 1983 when it was lost at sea. Simeon handled "the business end of things," i.e., the ship and dealing with the merchants, but he passed all the money over to Maud who handled it wisely. There were no mortgages taken on any of these ships. I expect that they had similar success in the fishery to that of Henry and Myrtle Hatcher and Simeon and Violet Blanche Butt.

Along the south and south-west coasts the same ownership pattern emerged. At Port aux Basques Michael and Hannah Maud McDonald jointly bought the *Allan Graham* in 1943. They obviously moved to North Sydney in 1950 since they transferred the registry of the ship to that port. Albert and Elsie Read of La Poile bought the *Coombes* in 1945 from Robert Allan Newman, merchant at Petites; the fate of that vessel cannot be determined from the records. We have already noted the Durnfords of Francois.

Before leaving the south-west coast, we should note that there was some modeling going on here. What was started by one couple became a model for many others. Joint ownership by couples made it easier to transfer full ownership to a surviving spouse; exclusive ownership may have created more difficulties. The advantage to the family of leaving everything to the wife is that no taxes were assessed in the event of the husband's death. There also existed in many Newfoundland communities a deeply ingrained superstition about making a will; it was either postponed until death from natural causes was imminent or it was regarded in superstitious lore as a prelude to death, i.e., bad luck. Simeon Butt, for example, registered all his property in his wife's name (ship, house, and so forth.) but did not make a will until after his wife died. The pattern of joint ownership by husband and wife became a common practice on the south-west coast, and I surmise that women's influence both initiated and spread the pattern throughout the region.

Owning Ships Jointly with Other Women

There are several cases of women owning ships jointly with other women, and they fall into two categories: two women own a ship with one or more men; two or more women jointly own a ship. There were only two instances of the first. Mary Howell and Jane Gould, widows and administratrixes to the estate of the late William Howell of Carbonear, were partners with William Willis Bemister when their 84-ton ship *Benjamin* was registered in 1839. The ship had been rebuilt before it was re-registered in 1839; it was totally lost in 1870 and it appears that both women retained ownership of their joint 32 shares during this period. As was seen earlier, Caroline Winsor, married woman, and Jane Winsor, spinster, were joint owners with four Winsor males at Aquaforte when the *Pelter* was registered in 1844.

The first instance of the second category is joint ownership between Mary and Ann Taylor, spinsters of Carbonear, when they registered their schooner *Mary & Ann* (note the name) in 1870. Each held 32 shares. The ship was lost near Copper Island, Fogo, on November 16, 1870, undoubtedly on the way back from the Labrador. Emmie Wilcox and Henrietta Roberts, married women of Brigus, bought the ship *Hyacinth* in 1902, and it was lost near Change Islands in August 1909. Edward Murphy, planter of St. John's, sold the *Bertha May* to Mary and Annie Murphy, married women, in 1905. Mary Murphy died in 1909, and Annie acquired full title to all shares. Annie took a mortgage on the ship in 1916 and it was discharged in 1919, at which time she sold the ship to Elizabeth Murphy, married woman.

In 1909 Jane and Ellen Rideout of Bonavista Bay, married women, bought the *Cecil Belle* from Edward F. Harvey, St. John's merchant, and they sold it to Job Bros. & Co. in 1915, after six years of use. Dorcas Phillips, widow, and Maggie Anstey, married woman, of Twillingate were appointed executresses for the estates of Martin and James Phillips when the two men died in January and May of 1927, respectively; they sold the ship *Mabel* after having owned it for only five months. The 56-ton schooner *Pandora* was bought by John and Josiah Yetman, fishermen of Harbour Grace, in 1906, and in 1909 they sold the vessel to Susannah and Minnie Yetman, married women. The women took a mortgage with John Rorke of Carbonear in 1910; the mortgage was transferred to John T. McRae, merchant at Harbour Grace, in 1913 and the mortgage was discharged in 1920. The vessel was sold as soon as the mortgage was discharged. Albert Hodder of Creston sold his ship *Ernie Loretta* to Lily Eliza and Sadie Pearl Hodder, married women of the same place, in 1958. They held 32 shares each. Lily Eliza was appointed the managing owner. In May 1960 Lily Eliza sold her 32 shares, and in May 1964 Sadie Pearl sold her shares.

In April 1937 Anna Mary Manuel and Myra Maud Bennett, married women of Corner Brook, bought the 101-ton ship *Dante* from Baine Johnston & Co. Ltd. and sold it in October 1939. Mary Florence Forward and Eleanor Tibbo of Grand Bank each owned 32 shares in the vessel *Chesley R.* for more than ten years; their husbands were business partners in the firm Forward and Tibbo.

Owning Ships Jointly with Men Who Are Not Their Husbands

Several of these types of cases were discussed in the first section on joint ownership, for example, Mary Parsons of Clown's Cove, Conception Bay. Mary Parsons, widow, jointly owned the ship *Venus* with Edward and John Parsons in 1818.

Mary Ann Stone, widow of Greenspond, owned 32 shares in the *Dash* when it was registered in 1867. George Allan, planter, owned the other 32 shares. Mary Ann was the widow of Benjamin Stone and married George Wright in 1868. George Allan was married to Rachel Stone, undoubtedly Benjamin's sister. Ada Mugford, married woman and later widow of Port de Grave, owned shares in the 43-ton *Tiger* from 1867 with William Mugford, Sr. (probably Ada's husband), William Mugford, Jr., Anthony Mugford and Elijah Mugford. William Sr. died in 1872. The three remaining males sold a total of 48 shares to Ada in 1887, and they held only 16 shares among them. In 1908 they sold a total of 32 shares to Abraham

Stone. This may well be a case of a woman owner with her husband and three sons. William Mugford, Sr., had two separate parcels of land at Port de Grave bequeathed to him alone and to him and his two sisters by his father's will of 1789.

Caroline Critchell, widow of Matthew of Belleoram, owned 16 shares in the *Marconi* in 1902 with her two sons, Uriah and James R. Critchell. Jane Collier, widow of Ship Cove, Bay d'Espoir, was a joint owner with Samuel Organ, fisherman and builder of the *Frances* when it was registered in 1887. Samuel Collier had owned shares in the ship when it was registered in 1869 so he had undoubtedly been Jane's husband. The fate of the vessel cannot be determined from the register. Mary Ann Brenton, wife of Charles Brenton of Belleoram, owned 32 shares in the *Bellona* in 1900; James Brenton, planter, owned the other 32 shares. I learned from interviews in Belleoram that Mary Ann was very involved in the business end of the fishing venture; she and Charles were the parents of Rev. Jacob Brenton of St. John's.

Mary Power, spinster of St. John's, bought 32 shares in the 115-ton *A. G. Eisnor* in 1922. John J. Duff held the other shares. Ellie Ann Hartigan, widow of Rencontre, bought the ship *Annie F. Hartigan*, along with William and Nicholas Hartigan, in 1922. The vessel was "wholly lost" on November 16, 1933, at Atlantic Cove, St. Paul's Island, in the Gulf of St. Lawrence, so it appears that Annie remained in the business with her partners for a total of 11 years. Annie Hartery, spinster of Cape Broyle, was an owner of a ship newly built by Michael Hartery, the 65-ton *Monica Hartery* along with Douglas, Martin and John Mulcahy, General Dealers at Cape Broyle in 1927. They sold the vessel to Bowring Bros. in 1929. From a personal diary at the Newfoundland Provincial Archives I was able to determine that the *Monica Hartery* was lost with all hands in January 1934, that Mrs. Monica Hartery died in 1936, and that Annie was married in 1944. There were several other women who owned ships as partners with men other than their husbands but those described previously give sufficient evidence that it was a viable ownership option for many women who wanted a hands-on involvement with the ship business.

Women Who May Have "Staked" Planters or Shipbuilders

This section may be a surmise on my part, but I am inclined to think that it may indeed have a basis in fact. It is quite possible that widows in the early years used their financial resources to keep their former husband's partners going in the fishing business. This may well have been true of Margaret McCarthy of Carbonear in 1823, of Olive Hederson of Brigus in the same year, of Mary Taylor of Carbonear in 1826, of Elizabeth Taylor of Carbonear in the same year, of Catherine Morris of Burin and St. John's in the early 1830s, of Ann Toque of Carbonear throughout the 1830s, of Elizabeth Henderson of Harbour Grace in 1837, and of Catherine Brown of Burin in 1844. It may also be true of Ann Giles of Carbonear in 1853, of Catherine Stoneman of Trinity and St. John's in 1858, of Mary Ann Stone of Greenspond in 1867, and of Jane Collier of Ship Cove, Bay D'Espoir in 1887.

A seemingly clearer case involved Elizabeth Freake, spinster of Joe Batt's Arm, in 1889. She was the owner of all 64 shares in the *Homeward* when it was built by her brother, Charles Freake of Birchy Bay. He was also the master of the vessel. No further information on the vessel is available from the register except that "it was broken up many years ago" and the registry was closed in 1945. My guess is that Elizabeth staked her brother in the building of the vessel and they shared the profits from the fishing venture. It is known that the Freakes moved from Joe Batt's Arm to Lewisporte, and on January 12, 1912, Elizabeth Freake of Lewisporte was granted one acre and nine perchas of land at Middle Mussel Bed Island, Burnt Bay, and that it abutted land owned by Henry Freake and Uriah Freake.

Gertrude Newman, spinster of Boyd's Cove, owned the ship *Margaret Newman* which had been built by her brother, Joseph Newman. Joseph had been a trader for at least four years before he built this ship. This seems to have been a partnership in which Gertrude was putting up the money, which she took in the form of a mortgage. Rosella Emberley, spinster of Creston, owned all shares in the *Rose and Blanche* which was built by William Emberley, possibly her brother, in 1928. Enola Fudge, of Belleoram, whose husband was Edward Owen Fudge, was a partner with George Yarn in the *Mary Hearn* in 1926. From interviews I did at Belleoram I would guess that Enola may have staked George Yarn.

Women as Executors and Administrators of Wills and Estates

There are several sub-categories to be dealt with in this section. It should be noted that in the early periods, women were consistently referred to in the clumsy anachronistic terms "executresses" and "administratrixes."

i. About 25 women in the ship registers were appointed to serve as executors or administrators along with one other man or with up to four other men, the latter in connection with very large estates. As many as 12 transactions were involved in settling some estates, including sale of ships, discharging mortgages and transfer of mortgages.

ii. More than 80 women were appointed as sole executors of wills and estates, which indicates both the trust that husbands had in their wives' business acumen, or fathers in their daughters', and the intimate involvement of the woman in the business affairs of the family.

iii. In two cases two women were appointed as the only executors of wills. Mary Howell and Jane Gould were appointed administrators of the estate of William Howell, merchant of Carbonear, in 1839, and they elected to remain in the business and did so for another 11 years until the vessel *Benjamin* was lost in 1850. The only other case involved Dorcas Phillips and Maggie Anstey to whom probate of the estates of Martin and James Phillips of Twillingate was granted in 1927 when James died in January and Martin in May of that same year. Dorcas was a widow; obviously either James or Martin had been her husband, and

Maggie was a married woman. They sold the ship *Mabel* in October of that year.

iv. More than 50 women between 1840 and 1983 who were appointed executors of a will or were granted Letters of Administration by the Supreme Court in order to settle the family estate sold their ship(s) within a year or two, some within a matter of days or weeks. For whatever reason these women were not interested in continuing the business – some because of age, some because their fortunes had already been attained, some because it was an opportune time to pass things over to other family members to continue the business, and some simply as a means of immediate survival. The reasons are just as likely to have been circumstantial as personal. I have chosen to deal only with a few from the earlier period because of possible family sensitivities.

Alice Elson was the widow of John Elson, who had been manager of Slade, Biddle & Co. of Carbonear. When John died in 1840 Alice was appointed administratrix to John's estate and she transferred by bill of sale to the trustees of the now insolvent company all John's rights in the two schooners *Tyro* and *Mary*. John Elson was a native of Exmouth, England, had been one of the principals in founding a literary circle at Carbonear and had also started a lending library in the town. Diana Butt's husband John, a planter at St. John's, died in 1847. His will appointed Diana executrix, and she sold the 42-ton schooner *Sultana* the same year. Mrs. Foley's (her first name is not available) husband Michael, a planter at St. John's, died in 1847, and his will appointed his wife executrix. She sold the 108-ton brigantine *Aspendus* within three months to the Lake brothers, "Booksellers in Falmouth in the County of Cornwall."

Julia Chancey's husband Thomas, of Carbonear, died without a will in 1848, and Julia was appointed administratrix of his "Estate and Effects." She sold his 16 shares in the 106-ton schooner *Adelaide* immediately. The Chanceys were a prominent merchant family. In 1848 their eldest daughter Sarah married Donald Bethune, Esq., "Sub-Collector of Her Majesty's Customs" at Carbonear, and in 1852 their daughter Cornelia married Thomas H. Newell of Boston.

Elizabeth Vallance Wakeham's husband, Robert Roberts Wakeham, died in 1848 and Elizabeth was appointed administratrix of the estate. Her husband had been a merchant, barrister-at-law, and clerk of the Legislative Assembly. She sold the brigantine *John Rachel* in 1848 and the brigantine *Ann Amelia* in 1849.

Mrs. Carter of Greenspond, widow of John Carter, Sr., who died in 1853, as his administratrix, sold the 66-ton brig *Britannia* the same year. The Carter family had been on Greenspond Island since the seventeenth century.

Most of the other cases in this category are quite routine and straightforward, but the probate process provides some interesting insights into merchant families. Elizabeth Munn of Harbour Grace was the widow of Hon. Robert S. Munn, merchant and member of the House of Assembly. Robert died on December 17, 1894, and on January 2, 1895, Elizabeth issued a certificate to sell the 114-ton schooner *Rose of Torridge*, "empowering Robert C. Ehlers to sell the ship at Bristol, England, for a sum not less than 750 pounds sterling within three months from date of certificate." The ship was sold to a British subject at Carnarvon, Cape Province, April 11, 1895. Robert Munn was a native of Scotland and had arrived in Newfoundland in 1851 with the firm of Punton & Munn. He married Elizabeth Munden, a native of Newfoundland. He died just a week after the crash of the Union Bank and his firm collapsed soon after his death. In 1906 Elizabeth became a mortgagee for a ship bought by Norman Munn of Harbour Grace.

When John Woodhouse Owen, merchant at Twillingate, died in 1902, his will, dated 1890, appointed his wife Louisa executrix. She sold the 25-ton *Sea Bride* in April 1903 and sold the northside and southside business premises to Henry J. Earle, John's partner for many years. Louisa then moved to Wilton, Mowbray, England, although it is believed that she was a native of Twillingate and that her maiden name was Janes. It has also been suggested that Louisa may have been from Terra Nova Birkdale in the county of Lancashire, England.

Robert Scott, merchant at Fogo, died in January 1913, and his widow Elizabeth F. Scott sold the 71-ton steam/sail vessel *Annie* in June of that year. Robert Scott had been born in Glasgow, Scotland, in 1835 and had run a fishery supply business and a passenger and freight service to and from Fogo Island.

The Hon. James Pitts, merchant at St. John's, died in 1914, and his will appointed his wife, Mary Pitts, and the Eastern Trust Co. Ltd. as executors. Mary sold her 31 shares in the 520-ton three-masted steam/sail ship *Ranger* in November 1915. The Pitts family had settled at Lance Cove, Bell Island, as early as the 1730s and later moved to St. John's. Another member of the family, James S. Pitts, carried on a large fishery supply business for many years.

Henrietta G. Harvey's husband John, of St. John's, died in January 1920, and Henrietta and Edward F. Harvey were appointed executors through John's will of 1918. They sold the 297-ton three-masted vessel *Sunset Glow* in June 1920, the *Gwennie Burdock* in June 1921, and the *Arkadian K.* in May 1922.

Robert G. Rendell, a St. John's merchant, died in 1929, and his will of 1926 had appointed Mary Herder, his sister, and the Royal Trust Co. as his executors. Over a period of 16 months, Mary Herder was involved in the sale of six ships, the discharge of four mortgages and the transfer of two other mortgages.

Maria (Buffett) Parsons was appointed one of three executors when her father, George Abraham Buffett, merchant of Grand Bank, died in 1929. Maria's mother was Julia Forsey. Maria's brothers were Aaron Buffett, also a merchant of Grand Bank, and Dr. Claude Buffett who lived and practiced in Hawaii. Maria had received a B.A. from Mount Allison University and had studied music there. She taught music in Grand Bank and later in St. John's. She was also choir leader at the Grand Bank United Church for many years. She died in February 1970.

v. And finally there were more than 40 women who inherited ships and elected to remain in the business. Some of these were discussed in the early period: Olive Hederson of Brigus, Elizabeth Taylor, Ann Toque, Mary Howell and Jane Gould of Carbonear, Margaret Blake and Elizabeth Vallance Wakeham of St. John's.

At a later time, when the famous fish-killer and inventor of the cod trap, W. H. Whiteley, died in 1903, his widow Louisa, as executor, did not sell her two ships quickly. She remained in the business for almost three years. When Alexander Lawrence, merchant at St. John's, died in 1906, his widow Jane bought the 43-ton *Glencoe* from the estate in 1908 and held on to it until 1917. She bought the *Emily* in 1908 and also the *Jim L.* in the same year and sold the latter ship in 1913 when her mortgage was discharged. She bought the 63-ton *Lewisport* in 1918 and sold it the same day. She was an active woman.

Mary Maria West of Bay L'Argent inherited her husband's 23-ton ship *Louie H.* when Thomas William West died in 1909; she did not sell the vessel until 1916. Elizabeth Martin of New Perlican remained in the business for almost three years after her husband William James Martin died in 1914. Her husband's partner was a trader; they sold the ship in 1917. When William Mayo of Marystown died without a will in 1918, Letters of Administration were granted to Martha E. Banfield (his sister?) of Garnish. She did not sell the vessel until 1927. Mark Guy, blacksmith of Catalina, died in 1924 without a will, and Letters of Administration were granted to Elizabeth Jane (Guy) House. She did not sell the ship until 1936, so presumably, it was being used in the family business.

The case of Jane Osmond of Moreton's Harbour is intriguing. Her husband, Dawe P. Osmond, merchant, died in April 1920, and she and Charles White, the other executor, sold the ship *Morton* in June 1921. But in the same month that her husband died, she bought 21 shares in the *Laberge* from Edward Roberts, mariner of St. John's, and did not sell these until December 1924. She appears to have sold these shares only because Roberts sold his 43 shares. We saw a similar pattern earlier with Sarah Foote, married woman of Grand Bank, who as an executor for her father's estate sold the shares in the inherited ship after three years and then bought shares subsequently in two other ships. Mitchie Ann Crosbie, as one of the executors for her husband's estate when he died in 1932, held on to the ship *Wabby* until 1936.

Ellen Kenny, widow of Fermeuse, held her ship, the *M. A. Kenny*, for 12 years after her husband died in 1937; she and her co-executor, John Joseph Kenny, did not sell it until 1949. Louis Sheaves, merchant at Channel, bought the 122-ton *Lucille M. Colp* from A. H. Murray & Co. in 1943. Louis died without a will in 1944, and his widow, Mary Wilmot Sheaves, inherited the ship and held on to it until it went ashore on a ledge off Frenchman's Head at the entrance to Humber Arm in May 1949. From interviews at Channel in 1995, I learned that Mary Wilmot Sheaves was born a Bishop at Bonavista, came to Channel as a school-teacher, and married Louis when she was 29 years of age. She gained two children from Louis' previous marriage and had nine children with him. Louis had little formal schooling. Mary was the bookkeeper of the firm in a real partnership. A family member remarked: "She run the business. Women looked after the books and the men worked." A big part of Louis' business was transporting fish to the mainland and bringing coal from North Sydney. He owned several ships over the years. Mary died in 1973 at age 77.

Women and Wills

Women were no more inclined to make wills about their own property than were men. Perhaps the same superstitious lore affected their attitudes towards making a will.

Only one instance was found in the records of a woman having made a will, and that was Sophie Knight of Grand Bank in 1926. Six women were found to have died without wills: Julia Tessier of St. John's in 1893; Margaret Stone of Rocky Brook, Trinity Bay, in 1918; Sarah Pittman of Merasheen in 1933; Emma Louise Foote of Grand Bank in 1935; and Eva King of Lamaline in 1936.

Designated as Co-partners in Trade

"Co-partners in Trade" was not a common term used routinely even for male partners in the ship registers. I did not encounter it more than a dozen times. So it seemed unusual to find it used, especially for women, in the earlier period.

There are three instances in the ship registers in which women were referred to with this designation. For two ships which they owned jointly in 1831 and 1833, the *Shamrock* and the *Calypso*, Catherine Morris and Patrick Ryan of Burin and St. John's were clearly designated co-partners in trade. Ann Toque and Simon Levi were stated to be co-partners in trade in 1831 when the ship *Elizabeth* was registered. When the *Elizabeth* was re-registered in 1839, Ann Toque and John Rorke were declared to be co-partners in trade.

When the ship *Sybil* was registered in 1852, Charlotte Colelough and 15 others were said to be co-partners in trade. The latter was a clear case of speculators investing in the future. John Haggaret was a master mariner and master of the *Sybil*, but the other partners in the venture were: James Beaton and Peter Ritchie, tailors; Hamilton Fleming, architect; James Wheatley and Edward Henson King Belchar, accountants; David Muir, silversmith; William Thuin, John Watt and Robert Murphy, carpenters; Patrick Wiley and George McLennan, coopers; Andrew Fairie, plumber; James Vey, watchmaker; John Kingwell, schoolmaster; and Charlotte Colelough, spinster. In the case of Julia Kelligrew, when the ship *Ann* was registered in 1853, though her husband William and his partner John Goodridge are clearly designated co-partners in trade, the fact that Julia held 22 shares in trust for her two sons implies that she was included as a co-partner.

The term "co-partners in trade" was quickly superceded in the records by use of the simpler terms "partners" or "owners." The implication of the earlier term was that, when it was used for women, it implied a more substantive role than merely putting down their money.

Managing Owners of Ships

All sole owners were indeed managing owners. We see this clearly in the case of Mary (Shepherd) Gosse of Spaniard's Bay as described by Eric Gosse in his book *The Settling of Spaniard's Bay*. Robert Gosse III built the ship *Victoria* in 1861 with the help of his six sons, and he died prematurely in 1863. His widow Mary "immediately took overall command, found the money to outfit for the fishery in 1864-65, and appointed her son Robert as captain and her eldest son John Charles as master of the cod-seines. With her crew of six sons Mary sailed aboard the *Victoria* during both voyages as cook and chief supervisor of these expeditions. In 1866 Mary stayed ashore, having sold her schooner to Robert, and the remaining members of the family dispersed."

Two other similar matriarchs at Spaniard's Bay were Sarah Barrett and Catherine Gosse. The earliest plantation in Spaniard's Bay was granted to Sarah Barrett in 1755, and there is no mention in the records of either inheritance or purchase. The property was already a well-developed fishing establishment at that time and consisted of two houses, three gardens, a meadow and two flakes. The third woman of note in Spaniard's Bay was "the redoubtable Catherine" Gosse, widow of Robert Gosse, Sr. Robert died in 1794 and Catherine continued to operate the family fishing enterprise and died in 1818 at the age of 92. This is probably the same Catherine Gosse who is listed in the 1805 *Conception Bay Plantation Book* as purchasing land at Bread and Cheese Cove from Thomas Mann for 30 pounds sterling (no date given) and leasing it to Gosse & Chancey for 31 years at 5 pounds sterling per annum.

It is no stretch of the imagination to characterize all three of these women, and many others like them, as managing owners of plantations, fishing ventures and ships. The ancient term of "ship's husband" is used rarely in the ship registers and is gradually replaced in the records with the term managing owner. Curiously the term "ship's husband" was found in the records as late as 1945, 1956 and 1959. The earliest use of the term managing owner that I was able to find was 1907 and then it was used in connection with a male owner. In the register of 1911, the registrar added a note: "If there are more owners than one, the Registrar is requested to distinguish the Managing Owner by the letters 'M.O.' against his name." Note the use of "his" with the implication that only men will be managing owners. The definition of the term managing owner is stated as "the person to whom the management of the vessel is entrusted by and on behalf of the owner."

It was not until the mid-1950s that the term managing owner came to be applied to women, and women soon went beyond that status for themselves and some began to appoint men, usually their husbands, as managing owners of their ships. This could not be done by the registrar until he received a letter from the owner stating that such a person was to be the "M.O." of her ship.

The first specific reference to a woman being appointed managing owner was that of Lillian Martha Hynes of Harbour Breton. She bought the ship *Rita M. Farrell* in 1954 and re-named it the *Elizabeth & Helen*. Two months later she informed the registrar that she was the managing owner. From an interview in Harbour Breton in 1994, I discovered that Lillian had been working for four years at St. Martha's Hospital in Antigonish, supervising the preparation of meals and that she married Sandy Hynes in December 1941 when Sandy was home on leave from the Merchant Marine. During his leave the ship on which he had been working was torpedoed; Sandy later received a DVA pension. Steve Hynes, Sandy's brother, was master of the *Elizabeth & Helen*. Lillian had only a Grade 8 education and Mr. Day helped her keep the books. She sold the ship in 1961 and the man to whom she sold had to take a mortgage of more than $5,000.

Lillian Frances Winsor, married woman of St. John's, informed the registrar of shipping that she was to be the managing owner of her ship, the *Sylvia Joyce*. She had bought the ship in 1953 and sold it late in 1955. Rosanna Cutler of Ramea advised the registrar in 1955 that she was the managing owner of the *Estelle Travers* prior to selling the ship; it looks as if this was information required by the registrar in order to conclude details of the sale. Emily Thornhill of Grand Bank appointed herself managing owner of the *Miss Pubnico* when she bought it in 1955; she sold it the next year.

When Jesse Howell, master mariner of St. John's, died in 1954, his will appointed his wife Annie as executor. She inherited his 32 shares in the 142-ton *Jennie Elizabeth* and continued as a partner with Thomas Hollett, a St. John's merchant. Under her own signature and that of Thomas Hollett, Annie was appointed managing owner. In 1958 she bought Hollett's 32 shares and informed the registrar that she was now the managing owner. In July 1959 the vessel went ashore at Little Pound Island Reef near Badger's Quay and became "a constructive total loss," according to advice received from the owner.

If a husband and wife owned a ship jointly and one of them was appointed managing owner, the signatures of both parties were required to be submitted to the registrar. For example, when Robert Allan Newman, merchant of Petites, sold the ship *Coombes* to Albert Read and Elsie Read of La Poile, fisherman and housewife, in 1957, the registrar's note reads: "Albert Read appointed managing owner. Advice under the hands of A. Read and Elsie Read received 18 November 1957." Likewise, when Frederick Clement Berkshire, fisherman of Spencer's Cove, Placentia Bay, sold 40 shares in the ship *Bertha Joyce* to Annie Northover Berkshire, married woman, and 14 shares to Raymond Burton Berkshire, mariner, and 10 shares to David Keith Berkshire, student, Annie was appointed managing owner and advice to the registrar required Annie's signature and those of the other two new owners.

Joseph Delurey, coasting master of Riverhead, St. Mary's Bay, and his wife, Laura Gertrude Delurey, were co-owners of the *Maggie Greene* when they bought the ship in 1952. In 1956 Laura took over 44 shares; Joseph Patrick Delurey, fireman, bought 5 shares, Harold Joseph Delurey, David Joseph Delurey and Joseph Delurey, Jr., all students, bought 5 shares each. Laura was appointed managing owner of the ship, and the advice to the registrar required her signature and those of the other four owners. We saw earlier that Chesley Yarn, merchant of Mose Ambrose, though he retained ownership of the ship *Wagaymack,* which he bought in 1956, appointed his wife Mary as managing owner of the vessel, and she used this ship in the pursuit of her separate business in English Harbour West. In 1956 Thomas Garland Ltd. of Gaultois acquired the ship *Teresa G.*, which had been built at Baddeck, Nova Scotia; the firm appointed Laura Garland manager in 1957; in 1966 the firm took a mortgage of $75,000 on the ship and it was discharged 14 months later. The ship was sold in 1968.

From 1955 to 1990 the ship registers contain the names of another 25 women who were named managing owners or managers of ships and fishery-related businesses. It is again testimony to the significant role that women were playing in the economy of the province.

Additionally, there are 25 instances in which women appointed men, usually their husbands, sometimes their fathers or a business partner, as managing owners of ships. Irene May Saunders of Cobb's Arm in 1959 appointed her husband Oliver as ship's husband of the *Fortunegull.* Sadie Pearl Hodder, married woman of Creston, advised the registrar that Philip Russell Hollett, merchant of Creston, was to be the managing owner of the *Ernie Loretta* after he bought 32 shares; she continued to hold the other 32.

Other cases are to be found in Appendix A.

Women and the Sale of Ships

There are three sub-categories to be examined here: husbands who sell ships to their wives and vice versa; married couples who sell their ships to other married couples; and women who sell their ships to other women.

There were 36 cases in which husbands sold their ships to wives, and 5 cases in which the wife sold to her husband. Emmanuel Stone of Rocky Brook, Trinity Bay, sold his newly built ship *Cumberland* to his wife, Margaret Stone, the same year that he built it, 1898.

The Hon. James J. Rogerson sold his ship *Orange Lily* to his wife, Isabella Rogerson, in 1900. He was described in the press as "a leading Temperance man, an earnest church worker and philanthropist." Isabella Whiteford was his second wife and she is described as "a singer of sweet strains" and "a sweet songstress," and by D. W. Prowse, the renowned Newfoundland historian, as "essentially a poet of the domestic affections." Prowse commends "…her humor, which is of the highest quality of true Irish pleasantry and fun." She published her first volume of poems in 1860 and her second in 1898. She died in 1905.

John and Josiah Yetman of Harbour Grace sold the 56-ton *Pandora* to Susannah and Minnie Yetman, married women of Harbour Grace, in 1909. John House of Gooseberry Island, planter of Bonavista Bay, sold the 54-ton *Lizzie H.* to Emma House, married woman of Gooseberry Island, in 1906. She sold the vessel in 1910. Charles Butt, planter of Spaniard's Bay, sold the *Fiona* to his wife, Emily Jane Butt, in 1909. According to the personal testimony of Eric Gosse when I interviewed him in 1995, Charles Butt ran a general retail store and was involved in the Labrador fishery; he was one of the great seal hunters of the period, a very enterprising man. He also had a great temper. As Eric Gosse remarked about Emily Jane: "She made herself felt. When Charles calmed down after one of his tirades, she told him what to do and made him think it was his idea." Eric Gosse wrote the story of the wreck of the *Fiona* on the Labrador coast and it was printed in the *Evening Telegram* about 1979. The crew "were able to put enough barrels and tierces around her that she floated off at high tide."

We have already looked at the sale of the *Chesley R.* by Felix Tibbo and Charles Forward of Grand Bank to their wives, Eleanor Tibbo and Mary Florence Forward, in 1934, and the sale of the *Agnes R.* to Violet Blanche Butt by her husband, Simeon Butt of Harbour Le Cou. Alphonsus Hickey of Petit Forte sold his ship *Meda* to his wife Bride Hickey in 1944, and she retained ownership until 1955 when she sold it to Alexander Haines, mariner of Pushthrough. Bride's birth name was Hayden and she was born at Petit Forte in 1883. She had a Grade 8 education, and she and Alphonsus had eight children. Family testimony affirms that she was in charge of the fish when it was brought ashore, assuring that it was properly washed and cured for market. She died in 1972.

There were five cases of wives selling their ships to their husbands. The first sale was between George V. Parsons, fisherman of Glovertown, and his wife, Susannah Parsons. Actually, George sold his ship, the *George V. Parsons,* to Susannah in 1918, and Susannah re-sold it to him in 1922. Elsie May Kendell sold her ship, the *Glimshire*, to William Kendell, mariner and her husband, in 1961, but she had appointed herself managing owner of the vessel when she bought it 17 months previously. The vessel ran ashore and broke up at Pushthrough in May 1967. Johannah Smith, married woman of Argentia, sold the *Margaret Ann Smith* to Anthony Smith, merchant of Argentia, in 1935. Anna Mary Manuel and Myra Maud Bennett, married women of Corner Brook, sold their ship, the *Dante,* in 1939 to Willis Manuel of Corner Brook, Gordon Scott Goodyear and Samuel James Anderson; I am reasonably certain that Willis was Anna Manuel's husband. Emily Thornhill, married woman of Grand Bank, sold the *Miss Pubnico,* which she had bought from Fortune Shipping Ltd. in 1955, to Ambrose Thornhill, oiler of Grand Bank, in 1956. This vessel went ashore and was a total loss at Fox Island near Ramea in 1958.

It is possible that a very few of the 36 cases may have had something to do with the protection of family assets, i.e., protecting them from merchants or creditors, but I would guess that these would be few indeed.

There are also two instances of married couples selling to other married couples. In 1947 Augustus and Phoebe Buckland, fisherman and married woman of Rose Blanche, sold the *Dolores Kaye* to Simeon Matthew Billard and Maud Billard of Harbour Le Cou, also fisherman and married woman. And in 1948 Gabriel and Bessie Billard, fisherman and married woman of Rose Blanche, sold the *Calvin Anne* to George Robert Durnford and May Durnford, merchant and married woman of Francois.

There were also six instances of women selling their ships to other women. Mrs. Anastassia McDonald of Salmonier sold her ship *Bonny Lass,* which Michael F. McDonald (presumably her husband) had rebuilt and enlarged in 1909, to Mrs. Ellen McDonald of Salmonier. As we saw earlier in the section on "Joint Ownership," Annie Murphy, married woman of St. John's, sold the ship *Bertha May* in 1919 to Elizabeth Murphy, also a married woman of St. John's. Lucy Roberts of Twillingate acquired all shares in the *Hattie A. Heckman* when her husband, Andrew Roberts, died without a will in 1926. She sold the vessel in 1927 to Mary A. T.

Roberts, spinster of Twillingate. Three years later Mary sold the ship to Ashbourne's Ltd. The *Jessie Beatrice*, which Ada Annie Petite, widow of Mose Ambrose, bought in July 1937, she sold to Gertrude Evans, married woman of Mose Ambrose in March 1938. Gertrude had a motor installed in the vessel and re-registered it. In 1953 Caroline Hannam, married woman of Rose Blanche, sold her ship *Wilehema* to Mary Elizabeth Hannam and Charles Hannam, married woman and fisherman of Rose Blanche. Meta Cutler, married woman of Ramea, appointed herself managing owner of the ship *Glimshire* when she bought it in 1959; she sold it in 1960 to Elsie May Kendell, married woman of Pushthrough.

Women, Ships and Mortgages

There are three sub-categories through which we may view women's roles regarding mortgages on ships: women as mortgagees for ships, women who took mortgages on their ships, and women who did not hold mortgages on their ships but the men to whom they sold had to take mortgages.

Twenty-six women in the Newfoundland ship registers were mortgagees, i.e., they provided mortgages for others to buy ships. Four of these women held two separate mortgages, and one was required after probate of her brother's will to discharge four mortgages and transfer two other mortgages. These cover the period 1895 to 1984.

The first mortgagee found in the records was Mary Layton, widow of Harbour Grace. She provided a mortgage of $800 at 6% interest for Henry W. Thomey to buy the 58-ton *Celerity* in 1895. Since she sold the vessel in June 1900, Thomey must have died or reneged on his loan. When Robert Fowlow, planter of Trinity, bought the ship *Etta May* in November 1896, Mrs. Selina Hepburn, married woman of St. John's, held the mortgage for him of $550 at 6% interest; the mortgage was discharged in October 1898. Nathaniel March, merchant of St. John's, bought the 317-ton *J. Percy Bertram* in 1900, and Anna March and Charlotte M. March, spinsters of St. John's (probably his sisters), were joint mortgagees to secure the sum of $6,000 at 4% interest. Their mortgage covered 24 shares.

Margaret Edens of St. John's held a mortgage on November 1906 of $1,050 at 6% interest for Levi Button, fisherman of St. John's, when he bought the 29-ton *C.W.J.* She discharged the mortgage in January 1907, though the mortgage agreement stated that it was not payable until May 1907. Button must have had a great fishing season. Elizabeth Munn, widow of Harbour Grace, held a mortgage of $4,000 at 6% interest in August 1906 for Norman Munn, merchant of the same place (probably her son), when he bought the 119-ton *Antoinette*. Though it was payable on August 1907, the mortgage was not discharged until 1912 when the vessel was sold. Annie, Cora and Octavia Grieve, spinsters of Edinburgh, Scotland, were the joint mortgagees for Thomas Crawford of Greenock, Scotland, for the amount of 1450 pounds sterling at 5% interest when he bought the 278-ton, three-masted *Attila* in 1912. Eight months later Crawford sold the vessel to Walter Baine Grieve, merchant of St. John's. There must have been a family connection here, and Crawford was probably buying the ship to deliver it to W. B. Grieve, merchant at St. John's.

John Chalker Crosbie, merchant of St. John's, bought the ship *Evelyn* in 1904. In March 1909, his wife Mitchie Ann Crosbie, held a mortgage on the ship for $6,000 at 6% interest, and in March 1910, she transferred the mortgage to Martha E. Bell, hotel proprietor of St. John's. This is yet another case where "Hotel Proprietor" is crossed out and "married woman" is written in by a later hand.

Sarah Vigus, married woman of Burin, held a mortgage of $3,263.74 at 7% interest when her husband James Vigus, merchant, bought the *Frances Mesquita* in 1916. The mortgage was payable on December 31, 1917; it was discharged on November 6, 1917. Sarah had come to Burin as a teacher and stayed on to marry James Vigus who died in 1918. In February 1919 Sarah Vigus became a mortgagee for the amount of $662 *without interest* when Robert

Wagg bought the *F. W. Mitchell*. The mortgage was to have been paid by November 1920. The vessel appears to have been seized by the High Sheriff under a writ of *fieri facias* on March 9, 1928, "over the shares of Robert Wagg." Four months later the vessel was sold to Gordon Hollett, accountant of Burin. Sarah had died on January 16, 1922, at age 63.

In 1931 Annie B. Blackwood, married woman of Wesleyville, held Mortgage B "to secure the amount due on Account current [amount not stated] at 3% per annum payable on demand" when Albert L. Blackwood bought the 108-ton *Flora S. Nickerson* in 1931. Ann Bishop, widow of Wesleyville, was the mortgagee for Mortgage C; in May 1939 she transferred this mortgage to Annie B. Blackwood. Albert Blackwood died in March 1938, and Annie was then a widow. It appears that Annie was still holding both mortgagees when the vessel was "totally lost" at the approach to Wesleyville Harbour on June 11, 1950.

In 1933 when Edmund Vardy, mariner of Hickman's Harbour, bought the 174-ton *Alcala*, Lillian Vardy, married woman of Hickman's Harbour, held Mortgage B for $4,000 with interest at 7% per annum; this mortgage covered 20 shares. The *Alcala* was a Lunenberg-built schooner and once raced the *Bluenose*, keeping up with the famous ship on the way out and losing the race by five minutes on the way back. Edmund Vardy bought the *Gertrude Jean*, a crude oil screw, in 1938 from Fanny Isabel Ryan Fiander, married woman of Trinity, and once again Lillian Vardy was the mortgagee for $4,000 with 3% interest. She discharged this mortgage in October 1941. From information received on Random Island, it is clear that Lillian was the business head of her husband Edward's firm and also extended mortgages to his brother Edmund. Lillian and Edward owned the *Flora Ann*, which had at one time been a rum-runner, and the *Silver City*, a steel-built boat. Lillian and Edward ran a general store, did freighting around the coast, and collected fish on the Labrador. She and Edward had three daughters, and one of their ships was named after two of the daughters, the *Ivy and Muriel*; the third daughter died very young.

Ship Thomas J. Hodder *(Maritime History Archive, MUN, pf 055-2-1032).*

Henry and Myrtle Hatcher of Big Bottom, Rose Blanche (Courtesy Mrs. Ann Parsons).

Ada Annie Petite
(Courtesy Mose Ambrose).

Ship Ada Annie Petite *(Author's Collection).*

Ship Fiona *(Maritime History Archive, MUN, pf 055-2-e066).*

Percy and Eleanor Carr on their honeymoon on the Restigouche River, New Brunswick, 1920 (Author's Collection).

Lydia (Foote) Hollett with husband, Wm. Hy. Hollett and daughter (Author's Collection).

Ship Maggie Stone *(Maritime History Archive, MUN, pf 001-1-b004a).*

Ship Maggie Stone *(Maritime History Archive, MUN, pf 001-1-b004b).*

Mary Harris, with her brother Chester Harris, MD, Toronto, c. 1921 (Author's Collection).

Ship Flora S. Nickerson *(Maritime History Archive, MUN, pf 001-u003a).*

Violet Blanche Butt (Simeon) with her grandson Melvin Parsons at Dalhousie University Convocation (Author's Collection).

Mrs. Eliza Pafford (Author's Collection).

The fourth woman who held two separate mortgages was Harriet Inkpen, formerly of Burin. Edward M. Hollett, merchant of Burin, bought the *Lottie M. Dunford* in 1932. In February 1935, Harriet Inkpen, spinster and then living at 46 St. Clair Gardens in Toronto, held a mortgage for $600 with interest at 6% per annum. At a later undetermined date she transferred the mortgage to the new owners, Joseph Green and Robert Gregory of Winterton, Trinity Bay, merchant and mariner. Perhaps Hollett had died. Also in February 1935, Harriet Inkpen provided a mortgage for $1,000 for the same Edward M. Hollett who had bought the *Betty Zane* in 1926. This vessel was lost at the entrance to St. Lawrence on December 6, 1938. What happened to the mortgage cannot be determined from the records.

Almost 50 women took mortgages when they bought their own ships. As early as January 1835 Simon Levi and Ann Toque of Carbonear took a mortgage on their ship *Alpha* with St. John's merchants George Richard Robinson, Thomas H. Brooking, John Bingley Garland and William Jeffrey Harvey. The mortgage was canceled in December of that same year, and Levi and Toque transferred all shares to the mortgagees "for a debt or debts." However they did not have a mortgage on their ship *Elizabeth* in 1831. Richard and Catherine Brown of Burin owned the ship *Catherine* in 1844, and in October 1847 Richard transferred his 48 shares "as a security for debt or debts" to Archibald Currie, merchant of St. John's, so obviously Currie held a mortgage for Richard. No mention is made of Catherine's 16 shares so she must have held on to them and possibly also a mortgage which she had with Currie. Margaret Blake, spinster of St. John's, had bought the ship *Triton* in February 1845, and when she sold it to John Connors, victualler of St. John's, in November 1847, it was by way of a "Deed of Mortgage," which means that he must have been her mortgagee. When Mary Foley Morris, spinster of St. John's, bought the ships *John & Mary*, *Nancy*, and *Relief* in 1842, she took mortgages on all three ships with James Niner Wood and James Douglas, merchants of St. John's.

Laura Simmons, married woman of Harry's Harbour, took a mortgage for $693.42 at 6% interest with George Knowling, merchant of St. John's, in 1895, one year after she registered the ship *Bonny Lass.* In June 1898 Knowling transferred this mortgage to James Norris, planter of Three Arms, Notre Dame Bay. In November of the same year, Norris sold the ship, which probably means that it was a repossession for unpaid debts. When Margaret Catherine Fitzgerald, matron of St. John's, bought the ship *Ada* in 1896, Robert J. Rendell, merchant, was her mortgagee for $500.

Martha Benson, married woman of Little Bay Islands, was one of four partners in the ship *St. Elmo* in 1898. Martha, Reuben, Arthur and Jonathan J. Benson took a mortgage in September 1898 "to secure sum due on account current with interest at 5%." James Strong and Richard Mursell of Little Bay Islands were the joint mortgagees. They took a second mortgage in June 1899 for $541.29 with Charles W. H. Tessier, merchant of St. John's. On June 12, 1914, Strong and Mursell discharged this mortgage; on May 29, 1913, Tessier discharged Mortgage B. They then took Mortgage C on June 5, 1903, though the amount was not given. The Bensons sold the vessel in June 1914.

Mary Murphy of St. John's owned the ship *Kersage* in 1899 and took a mortgage of $100 with interest at 7% from Charles W. J. Emerson, payable by November 1900. When Annie Murphy, married woman of St. John's, acquired full title to the *Bertha May* after Mary Murphy's death in 1909, she took a mortgage on the ship but not until May 1916 with Walter Baine Grieve, merchant, and this mortgage was discharged in January 1919. Other mortgages which women took up to the year 1983 were quite straightforward.

It may be of some interest to note that several women who owned ships without a mortgage sold their ships to men who had to take a mortgage. Mary Ann Hussey, married woman of Port de Grave, sold her 35-ton ship *Brothers* in 1907 to Abraham Richards of St. John's. She did not have a mortgage. Abraham took a mortgage. Amelia Jane Winsor, married woman of St. John's, bought the *Centenary* in 1924 and sold it in 1926 to Percy and Cecil Winsor, fishermen of Wesleyville. The men took a mortgage with A. E. Hickman Co. Ltd. Amelia Jane probably had connections to Wesleyville and may have been selling her ship to her sons.

When Harriet Chislett of Rose Blanche sold her shares in the ship *H. B. Chislett*, the buyer took a mortgage; Harriet did not have a mortgage. After using their ship *Minnie & Joan* for ten years in the fishery, Henry and Myrtle Hatcher of Rose Blanche sold the ship to their son Roy Gordon Hatcher, fisherman of Rencontre West, in 1956, and he took a mortgage with the Nickerson Outfitting Co. Ltd. of North Sydney.

Lillian Martha Hynes, married woman of Harbour Breton, sold the *Elizabeth & Helen* in 1961 to Walter Norman Penney, mariner of Little Bay Islands, and he took a mortgage with I. A. C. Ltd. for $5,446.54 including interest. When Jessie Bragg of Port Union sold the *Marie Yvonne* in 1964 to Walter Edmund Collins and Benjamin Garfield Collins, they took a mortgage; Jessie did not have a mortgage.

When Zena Leonard of Southern Harbour, Placentia Bay, inherited her husband's ship *Terry Maurice* on his death in 1977, she sold the ship immediately. They did not have a mortgage. John Carl Pelham of Pool's Cove, who bought the ship, took a mortgage. The same year, Florentine Hynes of Fox Island River, managing owner of the ship *Mary Evon*, inherited the ship when her husband died without a will. She sold it the same year to Augustus Joseph Hynes, fisherman, and he took a mortgage for $2,550 with interest at 3.5%, and Florentine held the mortgage.

Owning Ships and the Value of Shares

It is not possible to be very precise about the value of a ship or the value of one of the 64 shares in a ship because there are so many variables involved: the time period, age of the ship, condition of the ship, size of the ship, and so forth. Curiously, very little information on prices can be found in the ship registers. Some information may be gleaned from the amount of the mortgage the new owner took, but this could be equally imprecise because it may have been only a part of the total value, or other assets could have been included in the total mortgage.

Perhaps a few instances in giving the sale price of ships will help, using examples from Atlantic Canada and Quebec. In 1860 in Quebec a 970-ton ship was offered for sale for not less than 8000 pounds sterling, and a ship of 336 tons for not less than 2000 pounds sterling, and a third ship of 395 tons for not less than $14,000, so there was considerable variation, and the price undoubtedly depended on some of the factors mentioned previously. Capt. Allan MacPherson sold his 52-ton schooner in 1867 in Cape Breton for better than $10 per ton; the ship was 10 years old. Dame Henriette Paquet of Quebec sold her 32 shares in a one-year-old, 106-ton ship for $425 in 1875; that would make the shares worth just over $13 each.

In 1894 Elizabeth Munn of Harbour Grace sold her 114-ton schooner for $15 per ton; it was 19 years old. The 71-ton schooner *Minot Light* was sold by Thomas Hickman of Souris, Prince Edward Island, in 1889 for 560 pounds sterling and it was 10 years old. It had been built of white oak in Essex, Massachusetts, and it was advertised as "good for 30 years to come." Henry Evans of Northern Arm, in 1899, sold his two-year-old schooner of 38 tons, the *Cabot*, for $1,150, which is better than $30 per ton. The Horwood Lumber Co. built the 372-ton *Attainment* in 1917 for $60,000 and sold it almost immediately to Campbell & McKay for $69,420, collecting a bounty of $5,952. In 1920 they built the 149-ton *Nancy Lee* for $41,392. The cost of building the 347-ton tern schooner *E. P. Theriault* in Belliveau Cove, Nova Scotia, in 1919 was reckoned to be about $62,000; that would be about $179 per ton. These instances provide a rough context in which we can look at the kind of money women were dealing with in owning ships. To help us understand the context a little better, it should be noted that 50 British pounds in 1938 would be equivalent to $3,500 U. S. in 2006, and 50,000 pounds sterling in 1846 was equal to 2.5 million pounds sterling in 1890; that means, in the latter case, that in a mere 44 years there was an increase of 50 fold.

Margaret Yarn, widow of Mose Ambrose, made a very astute purchase in 1929 when she bought the 17-ton schooner *Bessie M.* from John Robert Petite for "$1.00 and other valuable considerations paid to me by Margaret Yarn, widow and managing owner, mercantile business of James Yarn." I saw the actual bill of sale in possession of Mary Yarn in 1995. What the "other valuable considerations" were is unknown. Annie Hickey, married woman of Harbour Breton, in 1936 purchased the 12-ton *Eleanor & Barbara* for $383 which was about $32 per ton.

From the amount of the mortgages on ships, some further information may be gleaned. When Sarah Jane Johnson, married woman of Little Catalina, bought the 15-ton *Lego* in 1900 she took a mortgage of $250. Susannah and Minnie Yetman took a mortgage of $600 when they bought the 56-ton *Pandora* in 1910. Kate Osmond, widow of Moreton's Harbour, held a mortgage of $5,000 on 14 shares in the 124-ton *Laberge* in 1922. And Elsie May Blackwood and her husband Job took a mortgage for $8,000 in 1947 when they purchased the *Mary H. Hirtle.* Whether these mortgages covered the total or partial value of the ship cannot be determined.

Occupations of Newfoundland Women

Apart from the self-designated status of widow, spinster and married woman, there are some interesting occupations given in the ship registers. As a variation on the status of married woman, in only three cases was the term "wife of" used, which is quite different, for example, from the ship registers of New Brunswick where the full names of husbands are given routinely.

Ann Jane Tuck, Anne Farrel and Ellen Roach were designated as traders, which meant that they were involved in coastal trading along the south coast in the 1880s. Gertrude Newman of Boyd's Cove is designated a spinster but she was almost certainly a trader as well. Abigail Horwood was designated a fisherman, but this was an error and she was, in fact, a merchant; her partner was a fisherman. Nina Osmond of Exploits, Burnt Islands, is called a spinster but she also was a merchant. Mary A. Rose of Harbour Breton was designated a merchant, and this was crossed out by a later hand and "married woman" was written in. Martha E. Bell of St. John's was called hotel proprietor and then this also was crossed out and "married woman" written in. Frances Miles was designated a planter; and lest we think this was exceptional, so was Mary Normor of Bay Roberts in 1814, according to Folio FO3, p.103 of Wills at the Public Archives; Mary Fowlow, widow of Trinity, in 1885, was also listed as a planter. This would also have been the status of Mary Ludevig of Ferryland and many of the independent women in the early Plantation Books.

At least four women were appointed as managing owner of a ship by their husbands: Mary Yarn of Mose Ambrose in 1956, Laura Garland of Gaultois in 1957, Bessie Goodyear of Carmanville in 1958 and Annie Northover Berkshire of Spencer's Cove, Placentia Bay, in 1959. As demonstrated earlier in "Women as Managing Owners," many women were in effect managing owners from the very early years, and from the 1960s onwards, it became quite common for women to appoint themselves as managing owners of ships.

Margaret Catherine Fitzgerald of St. John's in 1896 was called a matron, so she must have been a woman of status. Agnes H. Fitzgerald in 1919 was designated a stenographer. Una May Youden in 1956 was designated an accountant, as was also Mary Rosella Bradley the same year. At least five women were called "housewife": Clara Teresa Hardy, Elsie Read, Marion Emerson White, Lillian Martha Hynes, and Mabel Roberts. In two of these cases – Clara Hardy and Marion White – this designation was later "corrected" to "married woman." Bessie Wilhelmina Savory of Lewisporte was designated in 1975 as "Homemaker."

Marie S. Penny was called business manager in 1954; Rose Ting of St. Lawrence in 1986 was called Company Director of Eldorado Seafoods Ltd.; Vivian Wright of St. John's in 1990 was designated as both managing owner and businessperson. And Hazel Marie Stuckless of Gander in 1973 was also given the double designation of managing owner and housewife. Ruth Story of Story Bros. Ltd., Portugal Cove, was in 1975 a "clerk" who was appointed manager of the firm..

Emily Caines of Bartlett's Harbour inherited the ship *Freddie C* when her husband Sidney Joseph Caines died in 1970, and she sold it in 1972; she is clearly designated in the records as a widow. Amie Caines of the same place in 1968 was designated a fisherman and managing owner of the ship *Wedgeport Lad*; the ship was sold in 1972 also, the same year Emily sold her ship. It is impossible to determine if this was the same person.

Women Naming Ships After Themselves

It was common practice to name ships after women, usually a daughter or wife of the builder or the owner, so the small amount of vanity involved in a woman naming a ship after herself may be easily forgiven. It may in fact be viewed as understandable assertiveness and a public statement that women were fully involved in the business of owning and operating ships. Though there was an old belief that re-naming a ship was unlucky, it seems that women ignored the superstition.

There were at least 21 instances in Newfoundland in which women named ships after themselves in the period 1837 to 1953. Elizabeth Henderson, widow of Harbour Grace, owned the ship *Elizabeth* jointly with her son-in-law, Charles Simms of St. John's, in 1837. The ship had been built in Devon, England, in that same year and sold to Henderson and Simms at St. John's. The name was very likely assigned at the time of registration. Catherine Brown, widow of Burin, was joint owner with Richard Brown, builder and master of the ship *Catherine* when it was registered in 1844. The next case of this kind was not until 1862, when Catherine Cummins, widow of St. John's and joint owner with Peter Cummins, master mariner, registered the ship *Kate Cummins* at St. John's. In 1870, Mary and Ann Taylor, spinsters of Carbonear, registered the ship *Mary and Ann.*

When the Commercial Bank of Newfoundland crashed in 1894, bankrupting the Duder firm, Margaret Eliza Duder, wife of Edwin John Duder, bought the brigantine *Maggie*, obviously named after her, under the Liquidation Act of 1895. Mary Walsh's ship, the *Mary Sheehan*, may have been named in 1905 for her birth name. Theresa Stone of Catalina had William Johnson build a 20-ton ship for her in 1906 and she named it the *Theresa S.* When Mark Guy, blacksmith of Catalina, died in 1924, Letters of Administration were granted to Elizabeth Jane House (nee Guy), married woman, and very likely Mark's sister. The ship was named the *Lizzie Guy*, and Elizabeth House did not sell the ship until 12 years later.

We have already encountered Margaret Stone of Rocky Brook, wife of Emmanuel Stone. When their jointly owned ship was rebuilt and re-registered in 1925, it was named the *Maggie Stone.* Ellie Ann Hartigan, widow of Rencontre, in 1922 named her ship *Annie F. Hartigan*, and this name may have been a reference to her middle name and possibly her birth name. When William Mayo of Marystown died in 1918, Letters of Administration were granted to Martha E. Banfield, married woman of Garnish, and probably William's sister, but the Letters were not granted until 1927 when she sold the ship. Obviously Martha, and probably her husband, used the ship for the intervening nine years, and the ship was named *Martha E.* The name of Mary Frances Bishop's ship in 1927 was *Mary & Bride Bishop* so perhaps the ship at Salmonier was named after her and a sister.

Rosella Emberley's ship, the *Rose and Blanche*, in 1928 may have been carrying her name and the name of another family member. In 1933, Wilson Riggs, clerk, and Alice Maria Nash, married woman of McCallum, purchased and renamed their two-year old ship, the

Alice M. Nash. Margaret Brenton, married woman of Burin, in 1934 named her ship the *Margaret Brenton*. Eva King, married woman of Lamaline, in the same year named her ship *Eva King*.

Mary Vallis of Coombe's, Fortune Bay, bought the *R. J. Devereux* in 1941 and registered the ship under the name *Mary King*, undoubtedly her birth name. Alice Hatcher of Rose Blanche in 1946 named her ship *A. M. Hatcher*, possibly after herself. Also in 1946, Harriett Chislett of Rose Blanche named her ship *H. B. Chislett*, which may have been a reference to her own name. The ship in which Annie Marie Cheeseman of Rushoon held shares in 1950 was named *Annie Cheeseman*. When Nicholas Joseph Walsh, mariner of Fermeuse, died in 1953, Letters were granted to Margaret Walsh, widow. She then sold the ship *Maggie Walsh*, which bore her name. And finally, the ship that Elizabeth Joanne Hatcher of Rose Blanche inherited when her husband Roy died in 1984 was named *Elizabeth Joanne*. She was the managing owner of the ship and she may well have named it.

Ships, Irregularities, and Even Hints of Scandal

We have already encountered the feisty, proactive Newfoundland women who were unafraid to use the courts to redress wrongs done to them in that early society. There was a special fierceness in them when it came to protecting the family's assets and assuring its survival.

Mrs. Catherine Clements, widow of Ferryland in 1785, was an example of the feisty woman. When she was prevented by James Rouse, her neighbour, from using the landwash to wash her fish and the beach to dry it, she and three of her male servants pulled down the flake which Rouse had built on disputed land between the two. Catherine acknowledged that the land was not hers, that it had been part of the original Potsbury Plantation, but that she had used it undisturbed for many years and felt she had the right to continue to do so. She was still operating two boats and a skiff in the fishery. (A skiff was larger than a fishing boat, usually

35 to 40 ft. long, and was operated with sails, but required four men to row it, each with a paddle). Mr. Justice Carter informed the governor in court at Ferryland that the Clements family had purchased at least part of the original plantation. The court record breaks off in the middle of a sentence, saying that "it is their opinion the said Beach is neither of said Clements right but as the said Clements had dried her fish there before they had given it her she being last in possession…" The court obviously found in favour of Catherine.

In Trinity in 1753 there were 72 heads of families, 62 mistresses of families, 21 women servants and an unstated number of male servants. Servants were also called "dieters," i.e., those who stayed over from the Old Country for the winter and served "for their keep," i.e., their diet. During the working season, May to October, these servants were paid 20 pounds sterling each. Elizabeth Tite was one of these "mistresses of families," though there is no mention of children; she may in fact have been a spinster, though it is more likely that she was a widow. She had three English male servants and six Irish servants; she kept seven of these all winter. She had one boat, one train vat and one stage. She had a neighbor, Mrs. Mary Waterman, who was not faring as well. In that same year, Mary was ordered to pay her four creditors a total of 33 pounds sterling annually "until the debt was discharged." We are not told of the extent of her indebtedness. While her payments were kept up "she was not to be subject to distraint and could go on with the fishery." The full extent of her fishing operation is not known, but an inventory of her effects shows that she was in possession of "casting nets, caplin seine, cod seine, lance seine," so she probably had more than one boat and a similar number of servants to Elizabeth Tite, since the court must have determined that she was able to repay her debts and still make a living.

These kinds of cases can be multiplied over and over again from the records, and there were other kinds of "irregularities" in the early records. An interesting case is described in the Ferryland papers in a dispute between the Carter and Nunan families in the early to mid-1800s. Richard Nason left property to a spinster,

Anstice Gorman, and her illegitimate son, Thomas Nason, in his will of 1818. In 1828 Anstice was a servant and heiress of Dan Nason, Jr. In her will (date not given) she is described as "late of the Town of Youghal in the County of Cork, Spinster," and she left her property to her nephew John Nunan. Details are scanty in these papers, but it would appear that Anstice was the mistress of Richard Nason. The papers also indicate that "negro slaves" were kept in Ferryland during this period.

The ship that Elizabeth Henderson and her son-in-law bought in 1837 at St. John's may have been a case of seizure, since it was sold to them through "the Magistrate of the Admiralty Court of this Island bearing date of 19 May 1837." That kind of high-sounding language is not typical of the language in the ship registers. There are other cases of seizure as well. Ada Annie Petite, widow of Mose Ambrose, bought the *Charlie and Eric*, which had originally been named the *Mystery*. This vessel had been "seized from foreigners (US subjects) and ordered to be sold at public auction by the Exchequer Court of Canada, Nova Scotia Admiralty District." It was purchased from the Sheriff of the County of Halifax. Another case of what appears to be seizure is recorded in the case of the 16-ton *Grace and Evelyn*, which was owned by Jane Irene Paul, married woman of Fortune. Transaction 1 reads as follows: "John Cahill, Sheriff of Newfoundland, by order of the Supreme Court of Newfoundland in Admiralty dated 11 May 1938, sold 1 June 1938 to Philip Joseph Lewis of St. John's, barrister-at-law." Could this have been a case of smuggling from nearby St. Pierre?

The vessel *Night Train* had been off the registry for some reason prior to 1975, was restored to the registry in that year, sold to Harold Derber of Manchester, England, and then in Transaction 10 (the tenth ownership) it is stated: "Catherine Logan, under Power of the United States District Court for the Southern District of Florida, dated February 15, 1977 (as person from whom title is derived)..." Six months later the vessel was sold to Argus Shipping Co. Ltd. of the Cayman Islands. It appears that there were some irregularities involved here.

Mary Campbell, widow of St. John's, bought the *Swift Current* in 1894 from a Nova Scotia owner. Five months later the vessel was lost at Broad Cove Chapel, Nova Scotia; it was very likely involved in trading between Newfoundland and Nova Scotia. The registrar expressed his dissatisfaction with repeated attempts to get the Certificate of Registry from the owner, and finally states: "Application was made to Campbell & Smith 5 or 6 times for certificate of registry and could not get any satisfaction. Wrote to George Smith and after a time he came and produced the Halifax, Nova Scotia, certificate of Registry, May 6, 1896." Mary Campbell may not have been a good business manager in this case.

Edwin Duder of St. John's and Twillingate was at one time the owner of almost 200 vessels, sending more than 100 of these each year to the Labrador fishery and shipping fish to Europe and the West Indies in large carrier vessels. When Duder died in 1881 he passed his firm on to his son, Edwin John Duder, who built on the father's success. When the Commercial Bank crashed on December 10, 1894, Duder was unable to recover from the $600,000 indebtedness. He retired from the business and devoted his remaining years to managing the Carpasian Dairy Farm in St. John's; he died in 1918. (The fishery firm was revived in 1895 by a son-in-law, George J. Carter, and continued operating out of St. John's and Herring Neck). What is interesting is that on January 24, 1896, Margaret Eliza Duder, married woman of St. John's, bought the three-masted barquentine *Maggie* (undoubtedly named after her) "from the Commercial Bank of Newfoundland, under liquidation by act passed 7 January 1895, approved trustees: Jas. R. Fox, Maurice Teucton, Robt. L. Mare." This must have been one of the Duder carrier vessels. In 1904 Margaret Eliza sold 4 shares in the vessel to Edward Dingle, master mariner, and bought these back from his widow, Annie, when Edward died in 1905. Margaret Eliza continued in the ship business. In 1900 she bought the *Nimbus* and sold it two weeks later; obviously it was a money-maker. And in 1904 she bought the *May Flower* and held on to it for more than two years. The *Maggie* was lost at sea on March 11, 1912, and that seems to have marked the end of Margaret Eliza's involvement with ships. H. Y. Mott in his book

Newfoundland Men, published in 1894, describes Margaret Eliza with a flourish as "a lady whose charity and benevolence, as also her increasing work in the cause of philanthropy have given her a name, the remembrance of which will be handed down to successive generations for emulation." And P. K. Devine in *Ye Olde St. John's* states that Edwin J. Duder's wife was the daughter of Mr. Stead, merchant, and Miss Simms, and that she "was regarded as the most beautiful woman in the city, and is still living" (i.e., in 1936).

Bridget Mahoney, married woman of Bay Roberts, was the sole owner of the 26-ton schooner *Birch Hill* when it was totally lost at Batten's Rock, Labrador, on October 14, 1915. Another report says that the loss occurred during a storm at Stag Harbour Run, Fogo Island. The vessel had 240 quintals of fish on board, though another report says it was only 140 quintals. The Certificate of Registry was lost with the vessel, as were all the belongings of the crew members. The St. John's newspapers from October 16th to 23rd, and again in the following February, raised the spectre that this may have been a case of barratry, that is, the ship may have been deliberately cast away by the crew, which consisted of Bridget's father, husband and brothers, as well as other crew members. The matter went before the courts in a magisterial inquiry which lasted from February to May 1916 and ended in a judgment of barratry and a heavy sentence of two years imprisonment for each of six men, including "one old man." Since it was judged to be a case of barratry, the question then must be asked: Was Bridget knowingly a part of a possible criminal case? A newspaper report indicates that there was a crew of eight, "including a woman." This would almost certainly have been the cook, and most likely Bridget.

A well-known merchant of Notre Dame Bay in the 1910s and 1920s had a large fishing operation at Battle Harbour, Labrador, and used to trade along the northern Newfoundland and Labrador coast. He was described to me by a man who knew him as "a domineering tyrant and a price fixer." Though he had a wife, he also kept a mistress in a nearby town and took her with him for holidays in the United States. Both his wife and mistress were present in the house as the merchant lay dying. The wife stayed involved in the shipowning business; it is not known what happened to the mistress, except that she was well provided for, according to my witness. The witness further added a macabre postscript; the dogs dug up the merchant's grave after he had been buried.

Sarah Vigus, wife of merchant James Vigus of Burin, was in 1919 a mortgagee for a vessel, the *F. W. Mitchell*, which had been repossessed under a writ of *fieri facias* in 1916. In 1928, under a new owner, the High Sheriff of the Supreme Court of Newfoundland, again under a writ of *fieri facias*, seized the vessel a second time and sold it to a local merchant. Sarah apparently held the mortgage until her death in 1922. In a somewhat similar case, Frances Winsor, married woman of St. John's, bought the *Sylvia Joyce*, through Leonard T. Stick, a dealer at Bay Roberts, when the owner and builder of the vessel, Chesley Samuel Davis of Trinity, Bonavista Bay, went bankrupt in 1953.

When Norman Hatcher, owner of the *Topsail Girl Second*, died without a will in 1945, Letters of Administration were granted to his widow, Hattie Hatcher of Rose Blanche. She sold the vessel six months later. I was told by people on the south-west coast that Norman was killed in Nova Scotia and that later, a Nova Scotia man confessed to the killing.

As I visited homes and talked to people on the street in south coast towns, I was told stories of smuggling and rum-running, and that there were at least a few women who played a role in these activities. I had lived previously in two south coast towns and realized that smuggling from nearby St. Pierre was generally regarded as an innocent activity or even as an occasional challenge to outwit the authorities. Rum-running was a different activity, both highly illegal and dangerous, and sometimes with dire consequences. John P. Parker in *Sails of the Maritimes* describes the origin of this trade and its dangers: "In the dawn of the 1920's, however, a new trade developed which brought a measure of relief to some of the schooner owners. Prohibition in the United States, and later in Canada, created a thirsty population ready to pay any amount of money for 'real hard liquor.' As the domestic sources dried up and import restrictions were brought into effect, a new commercial enterprise in supplying ardent spirits, the 'rum runners' came into being." The two supply centres were the French islands of St. Pierre and Miquelon and the British and Foreign West Indies. Many shipowners quickly developed a new kind of ship charter to pick up liquor at the legal ports and carry it to the outside limits of American waters. Several Canadian vessels were involved in the trade and shipowners cautioned their captains to abide by the law and never to run the risk of losing the ship. There were serious risks involved in the trade and several vessels were captured when they either deliberately or inadvertently entered territorial waters. Captured vessels were either laid up or sold at auction.

I was told of at least one woman on the south coast whose vessel was involved in rum-running in the mid-1920s. The base for her vessel was a small town on the south coast; and when her husband bought a vessel during the same period he gave a Halifax address. He was described to me as "the king of the rum-runners."

Oddities and Noteworthy Women

The only clear reference I found in the ship registers to a woman having insurance on a ship was that of Mary Wilmot Sheaves, widow of Channel. When her vessel, the *Lucille M. Colp*, went ashore and was declared a total loss in May 1949, advice was received from both the owner and W. A. Munn Insurance Ltd. of St. John's. However, the fact that women insured their ships is demonstrated by the fact that when the 74-ton *Lady Green* was insured in 1929 for $2,500 in the Conception Bay Mutual Marine Insurance Co. Ltd., it was owned by Mrs. George Barbour of Newtown. Evelyn Ruth Hann, married woman of Wesleyville, became owner of the *Redwing Chief* in 1952, and the previous owner, Jesse Hann, had been an insurance underwriter who died in 1951. It is virtually certain that Evelyn Ruth would have continued the insurance on her ship. The ship registers are almost totally lacking in this kind of information, though I am convinced that many other women carried insurance on their ships.

Only one specific reference was found in the Newfoundland records to minors owning shares in a ship. Julia Kelligrew, wife of William of St. John's, held 11 shares each in trust for her two young sons, John William and William George, in the ship *Ann* in 1853. There are three references to students owning shares in ships, but it is impossible to determine their ages. The ship *Ariel* was sold to Samuel Piercey Evans, student of Grand Bank, in 1948. Joseph and Laura Gertrude Delurey of Riverhead, St. Mary's, sold 5 shares each to their three sons who are designated students in 1956. And Annie Northover Berkshire, married woman of Spencer's Cove, sold 10 shares in 1959 to David Keith Berkshire, student.

There are other curiosities. Three women owned ships that were involved in the seal fishery. One was the brigantine *William Stairs,* which was owned by Catherine Walsh, widow of St. John's. This vessel was "lost at Seal fishing in 1862." Catherine had owned the ship for five years. Charlotte Lee's ship *Mary Jane*, sometime between 1846 and 1862, was involved in the seal fishery, as was Emily Jane Butt's ship *Fiona* about 1909. Mrs. George Barbour's husband was a well-known sealing captain, and he may well have been master of the *Lady Green* (mentioned above) when it was insured by his wife in 1929. The *Lady Green* was likely a fourth woman's ship involved in the seal fishery.

Another curiosity: Why did Fanny Isabel Ryan Fiander, married woman of Trinity, buy the *Gertrude Jean*, a "crude oil screw" in 1937? It was undoubtedly built for someone else to operate, so this may be another example of a woman staking a man.

The only Labrador woman shipowner found in the ship registers was Joyce Allen, married woman of Rigolet. In 1978 she bought the *Sea Flower* from the Minister of Fisheries. She was designated managing owner of the ship and took two mortgages in the first two years and these were discharged a few years later.

There were other women shipowners whose names did not find their way into the ship registers but who have been reported to me as having been legitimate shipowners. Edith Brooks, whose husband Nathaniel was a merchant at Bay Bulls and possibly St. John's, owned ships, but the time period is not known. Jasper "Jap" Chaulk of Carmanville, owner of schooners for many years, died in 1937, and his wife inherited his ship *Mollie*, which had been built at Lunenburg and bought by him in the early 1930s. She was going to sell the ship but then decided that she would continue to oversee the operation by hiring a captain, which she did until her son Ross was able to take over the business in 1940. The ship ran ashore at Grates Cove in 1944.

Mrs. Ralph from St. John's owned the schooner *Linda Tibbo* in the 1940s; the ship sprung a leak and was lost with a load of salt in Harbour Buffett. I have heard also that Lillian Bouzane's grandmother or great-grandmother owned "a fleet of ships" and gave each of her sons a ship of his own when each became old enough to operate it. Her name was Philpott, and she may have lived in Coachman's Cove. This could not be verified from the ship registers, but Richard and John Philpott of Green Bay owned the ships *Mary Ann, Flora*, and *Lily Dale* in 1847, 1866 and 1891, respectively. Green Bay in those early years was taken to include Notre Dame Bay.

There were a few cases in which it was difficult to distinguish whether a name was male or female. Evelyn was judged to be a male name in the one instance where it was found. Also Celestine, Lorin, Wilga and Verdena. Their combinations with other names, as well as the owner's designation as "fisherman," leads me to believe all these were male. The one instance where Abigail was found (also spelled Abagail), designated clearly as "fisherman" was Abigail Horwood of Harbour Le Cou. This was an error in the register. Abigail was a merchant and her partner was a fisherman. For more details on Abigail see Appendix A under Joint Owners.

Although all of these women made significant contributions to the economic and social life of their communities and to the colony, and then province, of Newfoundland, there were several women worthy of special mention in the history of Newfoundland, such as Marie Smart Penny, Ada Annie Petite and Debbie Petite of the south coast. Charlotte Pratt Harris of Grand Bank was the sister of the famous poet E. J. Pratt and daughter of the Rev. John Pratt. Harriet Inkpen of Burin had already been honoured by the governor of Newfoundland for her bravery in trying to rescue three of her students from drowning before she went on to a distinguished career in Toronto as administrator of the St. Clair Gardens.

Myra Grimsley came to Newfoundland from England as a nurse in 1921, married Angus Bennett at Daniel's Harbour, and became "the nurse" along 200 miles of the west coast of the island, and was the subject of a book by H. Gordon Green, *Don't Have Your Baby in a Dory*. Several of the women of St. John's were noteworthy: Lady Mitchie Ann Crosbie, daughter of Josiah Manuel of Exploits, Burnt Islands, and wife of Sir John Crosbie; Henrietta Harvey, after whom the building that houses the Maritime History Archives at Memorial University is named; Isabella (Whiteford) Rogerson, the published poet; Louisa A. Whiteley (whose husband invented the cod trap); Margaret Eliza Duder; Elizabeth Vallance Wakeham; the Tessier women; Mary Herder; and Elizabeth Pitts.

And finally, "Big Allie" Murphy of King's Cove is worthy of special mention. According to Aidan Maloney, Alice Murphy "dominated the social life of King's Cove from 1840 to the 1890s." When the merchant James McBraire retired at King's Cove in 1821, he returned to Scotland with an accumulated fortune of 80,000 pounds sterling and left his son, John Joseph McBraire, in charge. After John Joseph died in 1839, Michael Murphy took over the business. Michael married Alice Denief of Fogo in 1841. "Big Allie" or "Big Mrs. Murphy" was 6 ft. 2 in. tall and "stout withal also." Michael was "a little Irishman," so they must have made an odd couple. Maloney writes: "When it is considered that after her husband's death in 1854 she continued to manage the business through good years, through fishery failures, and tragedies at sea it can be understood that the tributes to her ability, generosity and caring spirit are well deserved."

Women Who Went to Sea

Newfoundland females experienced the sea from the time they were infants; 1300 outports were scattered along the many miles of coastline and these had been originally located there to be near the fishing grounds and to wrest a living from the sea. In the early periods, when fishing was over for the season, many families left the outer islands or coastal villages to move into the interior, usually at the heads of bays where they had a ready supply of wood and food and areas ideal for hunting and winter fishing. The practice was called "winterhousing" and families erected temporary dwellings called tilts. Often the entire winter was occupied with the building of another ship for their own use or for sale.

As girls grew older, usually 14 or 15, they often went to the Labrador on the family fishing venture to act as cooks and domestics for the summer. Surplus daughters were sometimes hired out to other ships' crews as well.

Though it is known that some Newfoundland women went to sea with their captain husbands, I have been unable to find any diaries or journals from these voyages. Hattie Matthews of Burgeo, whose husband was William Matthews, sailed with him for about 10 years in the 1920s and 1930s when he was captain of the *General Laurie.* Hattie was the daughter of Captain George May, who had been a foreign-going master for many years. The *General Laurie* sank in the Mediterranean Sea, and presumably Hattie was aboard for that venture.

When Campbell & McKay's 372-ton, three-masted ship *Attainment* sank off Bear Cove, Digby, Nova Scotia, in a severe storm in April 1920, Captain Shepherd's wife was on board and, accompanied by a Portuguese sailor, she was sent ashore in a basket called a breeches buoy, as the sea was running "mountains high." Whether she ever recorded her memories of that event is unknown. All the crew were rescued in this fashion.

One fascinating diary does exist, and provided the basis for a recently published book by Bill Haigh, *Foote Prints Among the Kauri.* When the Foote and Gibbons families left Exploits, Burnt Islands, and St. John's in December 1864 to sail to New Zealand in the 132-ton brigantine *Clara,* which had been built at Prince Edward Island, there was on board a crew of 11 and 37 passengers, including 15 children. Mrs. M. S. Peace, a passenger, kept a diary of the six-month voyage, and Bill Haigh bases much of the early part of his book on Mrs. Peace's diary and quotes her poem "Farewell to Newfoundland" and her ode to the new baby born on board, "Little Clara's Song." Mrs. Peace described in detail every vessel sighting, a near disaster with a large steamer, rolling seas, shortage of drinking water at times, food spoiling in the hot weather, entertainments organized to dispel the boredom, singsongs, religious services, and the brief stopover in Capetown, where overindulging in liquid spirits inspired Mrs. Peace to write the following pledge: "We, the officers, passengers and seamen of the brigantine *Clara*, promise to abstain from all intoxicating drinks as a beverage until the present voyage is concluded," which she noted had only a limited effect. Running short of supplies, they stopped at Melbourne, Australia, where they were detained by officials for two whole months until they paid fines for six breaches of passenger regulations: "First, for not having sweet and wholesome provisions; second, for insufficient privy accommodation; third, for not having a supply of medical comforts; fourth, for insufficient hospital accommodation; fifth, for not producing a Master's list; sixth, for not having fitted lifebuoys." Then officials condemned the vessel's mast and said that it must be replaced and issued two more summonses: "First, for being two berths short; second, for not having the flour correctly stowed." Finally it was discovered that Captain Roper had not arranged for a Certificate of Clearance from an emigration agent in Newfoundland. It was agreed through a solicitor in Melbourne that all these additional expenses would be paid when the vessel was sold in Auckland. Several of the paying passengers left the ship at this point, intending to find an alternate way to New Zealand, and the *Clara* left on the last leg of her journey, arriving on June 30, 1865.

The Foote and Gibbons families had been sawmill operators at St. John's and at Peter's River, near the mouth of the Exploits River, since the early 1840s, and they were moving to New Zealand to continue that industry, which they did successfully for many years.

Newfoundland women also handled ships. Arthur Sidney Butt, in his book *Telling It As It Was*, affirms that the women of Flat Islands, Bonavista Bay, "handled boats almost as capably as men," and that "certain women acted as 'skippers' or 'skipperettes' when they went on berry-picking expeditions or when they shared off-the-island grass-cutting exchanges." Sarah Gosse Collins described her experiences in the late 1880s, as quoted in Eric Martin Gosse's book *The Settling of Spaniard's Bay*: "I sailed with father in the *Victoria* and with brother Bob (Gosse) in the *New Era*. We had some rough passages (to and from Labrador) but I was a good sailor. When they reefed I took my turn at the wheel and I could keep her into the wind with any of the boys. When I married John I never sailed again but every June I watched when the flags went to the mainpeak and they sailed away."

Surely one of the most incredible stories of women and the sea is that of young Anne Harvey of Isle aux Morts ("Island of the Dead") on the south-west coast of Newfoundland. On a stormy morning in July 1832 she was the first to hear the distress call of a ship breaking up on the treacherous rocky shoals. Anne, her father George, her younger brother, and Anne's Newfoundland dog went out in their dory to help. Her father realized that they must somehow get a rope aboard the ship so they could rig a breeches buoy in an attempt to rescue the people. The ship's lifeboats had all been smashed and the sea was full of floating debris as passengers and crew crowded together at the forecastle area. Anne suggested that her dog was such a strong swimmer that he could swim through the roiling sea and take the rope, tied around the dog's shoulders and middle, to the ship. He did, and the three Harveys were thus able to rescue 163 people and care for them until help arrived from outside the community a week later. Anne's role in the rescue, including her handling of the dory in the rough seas, was truly remarkable, but she had been fitted for the task by growing up on the sea and by the careful mentoring of her father. The Harvey family was awarded a medal for their bravery by the King of England. Many years later one of the Coast Guard vessels working the Newfoundland and Labrador coasts was named the *Anne Harvey* in recognition of the young woman's feat.

In Bay Roberts in the 1920s or 1930s there was a young woman whose surname was Batten who skippered a ship with her father as they brought coal from North Sydney to Newfoundland. Ann Miller of Fogo, whose husband was Paddy Miller, was skipper of a ship plying between her home and Botwood, probably in the 1950s. And more recently, Angela Butt was a 1998 graduate of the Marine Institute and served as captain of the *J & B* out of St. John's that summer. She was described in the *M. U. N. Gazette* as "possibly the first female schooner captain in the province's 500 year history."

Such was women's involvement in the family fishing venture that when the salt-fish trade began to disappear and fish caught on the fishing grounds could be taken directly to fish plants, it was a natural transition for women to stay in the boats and to be designated legitimately as fishermen, and later as fisherpersons. It is a cynical interpretation of the practice to say that women only did this at a later period so that they could collect unemployment insurance.

Later Newfoundland Women Shipowners

I checked the Newfoundland ship registers up to 1990, but I hesitate to give detailed information about some of the later women ship-owners for perceived reasons of personal sensitivities. Both in this chapter and in Appendix A, I have used the 1960s as a cut-off date in providing details of all transactions. It may, however, be of interest to list the names of later owners, with minimum details. These are a mixture of sole owners, joint owners, mortgagees and executrixes. The vessels are a mixture of fishing ships, working ships, yachts and pleasure boats. The abbreviation Mort. in the last column indicates that the woman was a mortgagee for the ship. The abbreviation M.O. is for managing owner.

Woman	Place	Occupation
Phyllis Marie Hillyard	Cupids	Married & M.O.
Joan Rayleen Sharpe	St. John's	Married
Helen Louise Pollock	Lewisporte	Housewife & M.O.
Susan Mary Petursson	St. John's	Married
Edna Ackerman	Trinity, B.B.	Married
Gertrude Kelly	Beau Bois, P.B.	Widow
Hazel Marie Stuckless	Gander	Housewife & M.O.
Eva Wareham	St. John's	Widow
Ruth Story	Portugal Cove	Clerk/Manager
Bessie Wilhelmina Savory	Lewisporte	Homemaker
Zena Leonard	Southern Hbr.	Widow
Florentine Hynes	Fox Island River	Widow & M.O.
Marjorie Brown	Baie Vert	Widow
Mabel Roberts	Springdale	Housewife
Joan Allison Herder	Topsail	Widow
Joan Allison Herder	Topsail	Widow
Maisie Herritt	Sandyville	Married
Elizabeth Joanne Hatcher	Rose Blanche	Married & M.O.
Mamie Weir	Little Bay Islands	Widow
Lenora Fagan	St. John's	Widow
Fannie Roberta Cole	Winterton	Widow
Judith Collins	St. John's	Married
Joan Verna Petten	Port de Grave	Married & M.O.
Rose Ting (or Ling)	St. Lawrence	Company Director
Rose Ting	St. Lawrence	Company Director
Rose Ting	St. Lawrence	Company Director
Rose Ting	St. Lawrence	Company Director
Vivian Wright	St. John's	Businessperson & M.O.

Date	Ship	Status
1968	*Pea Jay*	Sole
1969-72	*Cormorant*	Mort.
1970-	*Aigle de Mer*	Sole
1971-73	*Rusty Scott*	Sole
1971-	*Bonny Girl*	Sole
1972	*Mortier Bay*	Executrix
1973-75	*Miss Sheila Marie*	Sole
1974	*Barry & Carol*	Executrix
1969-72	*North Bay*	Sole
1975-78	*Flying Swan II*	Sole
1977	*Terry Maurice*	Executrix
1977	*Mary Evon*	Executrix
1978-79	*Elessia Elizabeth*	Executrix
1979	*Chris & Pete*	Executrix
1980	*Raybeu*	Executrix
1980	*Beothic II*	Executrix
1980	*Devon John*	Joint
1984	*Elizabeth Joanne*	Executrix
1983-87	*Miss Green Bay*	Executrix
1983	*Makkovik*	Executrix
1984-	*Judy Janet*	Mort.
1985-86	*Albatross V*	Joint
1985	*Sherry Roxanne*	Sole
1986-89	*Sandra & Carolann*	Sole
1986-89	*Cape La Hune*	Sole
1986-89	*Ashuanipi*	Sole
1986-89	*Nina Maria*	Sole
1990-	*Manta Ray*	Sole

CHAPTER FOUR

Summary and Conclusions

This book takes us back to a time when women, battling against almost insurmountable odds, carved out a place for themselves in previously unfamiliar territory.

How many women shipowners were there, and how do their numbers compare to those of male owners? In the microfiche edition of the ship registers to which I have referred several times, the listing covers the years from 1820 to 1936, i.e., 116 years. In 1820, 84 vessels were registered, and for the next three years the average number of registrations per year was about 40. Vessel registration reached a high of 243 in 1825 and then averaged about 95 vessels per year over the next five years. The first woman made an appearance in the listings in 1818 (Mary Parsons, widow of Clown's Cove), though Eric Gosse affirmed that Margaret and George McCarthy of Carbonear, who re-registered their ship *George* in 1823, were joint owners of that vessel as early as 1804.

A total of 97 women shipowners were found in the microfiche edition of the ship registers for the period 1820 to 1936. This number compares with a total of 16,629 owners for the period. In this total number we must consider that in many instances the names of five or six men (and even more) appear as joint owners of a particular ship, often the shipbuilder and his sons or a group of brothers. However, when the subsequent transactions to the original registrations were checked, an additional 173 women were found, making a total of 270 women shipowners. By searching the records from 1937 to 1990, another 138 women shipowners were found, for a total of 408. As is evident from this study, several other women boat owners and shipowners were identified from various written and oral sources. So we are looking at a Newfoundland sample of between 500 and 600 women.

In some of the very early census records, widows owning boats and plantations accounted for roughly 10% of the population. That seems unusually high, and the statistics above do not approach that figure. Again, by way of comparison, the *Conception Bay Plantation Book* of 1805 lists about 147 women property owners; these women had some kind of connection with 156 of the 1062 plantations or properties. However, not all of these women were boat owners and involved in fishing ventures. These were mostly waterfront properties but some were being leased to other tenants by women owners. There are several references to "on waterside," "by the seaside," "beach," "wharf," and so forth. There are specific references to flakes on 45 of the plantations, to stages on 39 and to "stores" on several. Some of the entries lack any kind of specific information, and at least one was not a waterfront property since it was said to be surrounded by woods on all four sides. It is impossible to be certain how many of these women were actually involved in the fishery, and there is no point in guessing.

And now, back to the ship registers for 1820 to 1990. When broken down into 20-year periods from 1820 onward, we find the following numbers of women shipowners:

1818	1
1820-1839	19
1840-1859	20
1860-1879	10
1880-1899	41
1900-1919	81

1920-1939	121
1940-1959	81
1960-1979	23
1980-1990	11
Grand Total	**408**

Again, it must be said that this number covers women in the ship registers only. There is a scarcity of factual data from the earlier periods and a lack of precise documentation in all too many instances. Despite these problems, can we now say that the objectives of this research have been "proven" or substantiated in a satisfactory way when we consider the evidence presented? The research set out to show that:

i. **Women were involved in a substantial way in the early economy and society of Newfoundland and this involvement included ownership of boats, ships and waterfront properties associated with the fishery.**

   This is probably the easiest of the objectives to substantiate. There is clear evidence in court records, in the plantation books, in the early census records, in the ship registers, and in the general literature that women's role as owners was substantial. We do not need to belabour this point.

ii. **It was real ownership and we are dealing here with real money.**

   Obviously, there were many cases in which registration of a ship in a woman's name, and particularly in a wife's name, was a paper transaction. No money changed hands in this type of transaction. But the intent was sincere and honest, as we saw with Henry and Myrtle Hatcher of Rose Blanche, and with Simeon and Violet Blanche Butt of Harbour Le Cou. The intent was that, should the husband die, the woman and her children were provided for, and other members of the family were limited in their claims on the estate.

A widow who continued to partner with her deceased husband's male partners was risking her livelihood, often on dangerous ventures. A spinster who bought a ship or ships on her own was risking her fortune, even if it was inherited money.

The money advanced by women mortgagees was real money. Many of these women had their own money, were receiving shares from a family business, or were involved in business transactions independent of their husbands. We saw how this worked at Grand Bank between the husband and wife team of Charles Forward and Mary Florence Forward. Women who advanced large mortgages were unafraid to take the attendant risks.

At Rose Blanche in 1994, I was told by the former owner of a long-liner that not one of his crew members earned less than $80,000 a year before the cod moratorium was imposed in 1992. A young woman in Harbour Le Cou said that before the cod moratorium, young people graduating from high school were given a car or a truck as a graduation present. She had worked for years in the fish plant for good wages, and suddenly there was no work. In a few short months they had gone from prosperity to near-poverty. When the fish plant closed, this young woman was trained in Port aux Basques as a hairdresser, but suddenly there was such a surplus of hairdressers that there was no work for most of those recently trained. All the new houses in communities along the south-west coast, like the houses with elevators along the Northern Peninsula, were built when the fishery was at its peak. So, yes there was good money being made in the fishery and women had a hand in using and saving it.

iii) Although it was real ownership, our society has collectively forgotten that there was a time in our maritime history when women's substantial roles were being played out.

Until very recently, even maritime historians seemed largely unaware of the significant number of women who owned

> ships and of the significant role these women played in our early economy and society. At the same time that British public figures and parliamentarians were fighting for clear property rights for married women, several married women in Newfoundland and the Maritime provinces of Canada were already being listed in the ship registers as sole owners and joint owners. Widows and spinsters owned ships and shares in ships at least from the 1700s, and many owned boats and plantations long before that.

It is a tragic commentary on our history that we have forgotten these facts. This is "history forgotten." Since I undertook this research project in 1993, I have had occasion to give public lectures and field questions on the subject and people have uniformly been surprised and unbelieving. When I found the first four women shipowners listed in the microfiche edition of the ship registers in 1992, my first reaction was "This must be an error. I will note it and check it out later." Others feel similarly, I am sure.

Perhaps one of the reasons why we have forgotten this history (as we shall see in the next point) is that women stepped into roles of leadership as needed (for example, following the death of a husband) and then stepped back out of those roles as circumstances improved (for example, a prosperous voyage, sons taking over the business, remarriage). Women were better at multi-tasking than men (I think); a woman who remained at home when her husband and his crew went to sea still had to care for the family, look after the kitchen gardens and animals, organize and superviseall work on the property, help to cure the fish when it was brought home, and so forth. So, what was one more job? There is no better confirmation of this point than Hilda Chaulk Murray's book *More Than Fifty Percent: Woman's Life in a Newfoundland Outport, 1900-1950*. Sean Cadigan confirms this observation in his thesis.

The real history of this situation needs to be restored or reconstructed, and we must never allow our children to forget what many women accomplished in the past, often under the most trying of circumstances. We must move past the comment that this was exceptional, understood in the sense that it was occasional, sporadic, and not to be taken seriously. People who make such assertions find it easier to forget the past.

As a rough parallel, I quote an experience I had with Robert Evans in 1955. As I sat down with the last of the great Evans shipbuilders, I asked him to give me the names of the vessels which the Evanses had built. Starting from 1870, about six or eight years before he was born, he gave me the names of practically every ship the family had built, the order in which they were built, their tonnage, their fate, and their association with events of the time and how this governed the names they were given. This information was all confirmed later by my research. Though he gave me the names of the seven ships he had built at Northern Arm from 1901 to 1909, he did not mention the two three-masters he had built: the 372-ton *Attainment* at Thwart Island in 1917 and the 149-ton *Nancy Lee* at Campbellton in 1920. These omissions would have been "history forgotten" had it not been for my conversations with him in 1955. And the amazing thing is that he thought that as a shipbuilder he was doing nothing remarkable. I am sure that he would have said, "But lots of other people were doing the exact same thing." Is that, I wonder, why we have forgotten women's role in connection with ownership of boats, ships, and waterfront properties associated with the fishery? They were simply doing what a lot of others were doing, doing what needed to be done. This would be in accord with Mary Yarn's comment to me in 1994 when I told her I had come to visit her because she had been a shipowner. She said, "No, I never owned a ship." I mentioned the name *Wagaymack* and she said, "Oh, that was a boat. I thought you meant a *big* ship." But the *Wagaymack* was a decked vessel and was therefore classified as a ship and required to be registered. It is a sad commentary that these kinds of things may seem to have been less remarkable because they were being done by women. No one at the time bothered to record such happenings.

iv. **Ownership was a natural activity and it evolved out of situations of necessity and opportunity.**

Perhaps the best example of natural ownership is described by Eric Gosse in his book *The Settling of Spaniard's Bay*. When Robert Gosse III died in 1863, he left his ship *Victoria* to his widow Mary. "She immediately took overall command" and for the next two years, ostensibly as the cook, she fished with her six grown sons on the French Shore. One of her first acts with her sons was to rescue the crew and passengers of a ship which had gone aground at Cat Harbour (Lumsden). After two years at the fishery Mary ran the operation from the shore and sent her young daughter as cook aboard the *Victoria*.

This was the pattern that women followed. When a husband died prematurely, often drowned at sea, the wife became the skipper of the fishing venture, hiring the crew, outfitting the ship, taking charge of curing the fish, dealing with the merchants, and sharing the profits and the risks. Women had certain advantages: they were the experts at curing fish, they generally had more schooling than their husbands, and they had a fierce determination to ensure that the family survived as an economic unit. For many women it happened that naturally. Those who survived were durable women indeed.

v) **Women possessed the requisite skills to be real partners in business ventures.**

It has been clearly demonstrated from the literature and from oral sources that women were the real experts at "curing" or "dressing" fish. Whether it was in a small family operation or the kind of commercial venture that was undertaken through large business firms as in Grand Bank, it was women who usually took the main responsibility of curing the fish when it was brought home from the fishing grounds and thrown up on the stages or hauled to the drying beaches.

Women usually had more schooling than men and often kept the family's accounts, acting as bookkeepers and being in charge of the money. I have quoted many cases to substantiate this. Taken from the book *Never Done: Three Centuries of Women's Work in Canada*, this comment is very telling: "But women also had an important work role in family businesses in early Canada. Although only a small portion of the population was engaged in trade, among the group most of the married women took part as permanent partners or replacements when their husbands were away. This situation was so common that women in business were an accepted part of the culture of New France, as they were in France itself." What was true of France and New France was correspondingly true of Great Britain and English Canada.

Women's work on the family property complemented the work of their men, whether it was raising children, keeping kitchen gardens, caring for animals, making hay, or taking leadership roles in the community.

vi) Older women modeled shipowning and property ownership for younger women.

Several examples of this modeling were cited throughout this study. While modeling seems to have reached almost an art form on the south-west coast, it seems evident in other places as well. Perhaps one of the key figures on the south-west coast was Harriet Billard of Rose Blanche, wife of Samuel, who owned the ship *Ethel & Albert* from 1932 to 1939 and had an engine installed in the vessel in 1936. She may in fact have modeled shipowning for other women along the south-west coast. Maud Billard of nearby Harbour Le Cou, Harriet's niece, became a shipowner for most of the years between 1947 and 1983. Maud jointly owned the *Dolores Kaye* with her husband, Simeon Matthew Billard, from 1947 to 1953, and the *Austin & Dianne* from 1963 to 1983. Their partnership was described to me in 1994 in this way: "Simeon handled the business and Maud handled the money. Simeon passed all the money over to Maud." The example of such women would have been felt in the entire south-west coast area. Women who sold ships to other women, or couples who sold to other couples, were consciously modeling ownership patterns.

Women from 130 Newfoundland communities owned ships in the period under study, from approximately 1800 to the 1960s, according to the ship registers. Apart from discrete areas like the south-west coast where women's ownership would have become known by word of mouth, many women accompanied their husbands to St. John's by ship for at least an annual outing. This kind of information would have been spread as women went "gamming" (visiting) aboard other ships in St. John's harbour, and as they met while shopping on Water Street and Duckworth Street and especially in the milliners' shops.

It is clear that older women modeled shipowning for younger women, whether consciously or not. It may well be that early women thought they were not doing anything extraordinary, that they were, in fact, prepared to do anything both to prove and to enhance their natural abilities. It was the survival of the family or small community group as an economic unit that was their primary motivation.

Women modeled leadership for other women in church and community activities throughout Newfoundland. The Wesleyan-Methodist Church was particularly adept at this, mainly because of Susannah Wesley's influence on her son John Wesley, founder of the movement. Women were trained along with men as class leaders by Wesleyan-Methodist clergy. It was a rigorous training and required reading, study, record-keeping and counseling, and women were given responsibility to teach and supervise classes of women, and when there were insufficient male class leaders, men were enrolled in a woman's class. These same skills were used in their teaching of children and youth in Sunday School classes. Hannah Eliza Evans, a class leader and Sunday School teacher at Northern Arm, also operated a branch store for the George J. Carter fishery firm at Herring Neck; the skills she learned in the class meetings were transferrable to the business world. In the same way that the religious revivals of the early nineteenth century in the United States helped to encourage strong female activism in the suffragist, abolitionist and temperance movements, so the early Methodist revivals in Newfoundland (strongly opposed and threatened by the "mother church," the Church of England), with their deep emotional dimension, gave women both a strong confidence in their own abilities and the skills to take active leadership roles in their communities.

A further conclusion that can be drawn from the evidence presented is that it is time to dispense with the near-myth of the passive woman in this early society. There are those who persist in portraying women as weak and powerless in early society, perhaps in some cases because it has become a convenient tool in the present to help redress the wrongs done to women through the centuries. However, passive women are not the picture that emerges from the court records or census records or the plantation books of Newfoundland, among other sources. We cannot discount the courageous, feisty, bold women who stand out in these records because they would not suffer wrongs to be done to them or their families, and they used every means to ensure that their cases were heard and action taken. They did not always win the individual case, but their voices were heard, and word spread.

There are at least four patterns of relationship between a woman and her partner(s) that emerge in the records. It may be instructive to look at these patterns as a way of determining whether women played an active or passive role in ownership matters. I would characterize them in this way:

a. **Women were subordinate in the relationship.**

A good example of this pattern is the beach women of Grand Bank. These women were gainfully employed; they worked for the large business firms, not for their own family business. They worked in teams, a boss woman being in charge of nine other women. They were the experts at curing fish. When the fish was brought to the beach from the harbour, these groups of women were responsible for the finished product that was then taken back to the harbour and shipped to Europe or the Indies. The last word on the curing was never the woman's. There was a man (a "culler") appointed to inspect the fish and to give the final word on the product. I have never read of an instance in which the boss woman was promoted to the job of "culler." The women, important as their role was, were in a subordinate relationship.

Other cases of women in subordinate roles may be those whose husbands registered their ships in the wife's name in order to escape indebtedness to the merchant. It is impossible to identify these with any degree of certainty, but one case may have been Susannah Parsons of Glovertown, who from 1918 to 1922 was technically the owner of the ship *George V. Parsons*, named after her husband. I have it from personal testimony that Susannah was ill for much of her life and was in no way involved in the fishing venture. Susannah was originally from Carbonear and may have come to Glovertown as a teacher. If that is not so, she may still have had sufficient schooling to keep the books and handle the money. In all such cases we must remember the viability of the family as the primary economic unit; everything was subordinate to the family's survival.

b. **A woman was subordinate in a real partnership.**

This pattern would have been true of Myrtle and Henry Hatcher in Rose Blanche. And from my visit with them in 1994, I don't think Myrtle would have had it any other way. It was a good and fair partnership. It appears that Henry was ensuring his wife's financial security, perhaps against a possible claim by one of the children from his first marriage. Myrtle was a smart, savvy woman, but it was the family's survival that was uppermost in her mind, and I think she would have deferred to her husband on most issues.

Merchants' wives who owned a few shares in the family firm's ships and who played a supporting role may have belonged in this category. Emma Wheeler of Frenchman's Cove, Bay of Islands, was a joint owner of the ship *Spring Bird* with her husband Symeon in 1896. Symeon was a lobster packer. This was a joint venture, with Emma almost surely playing a supporting role. Eleanor Tibbo and Mary Florence Forward of Grand Bank were real partners with their husbands in the Forward & Tibbo firm, though most of their activity was from behind the scenes.

c. **A woman was in an equal partnership with her partner(s).**

It may well be that Simeon and Maud Billard of Harbour Le Cou had such a partnership, where "he handled the business and she handled the money." Each had complementary skills and they were deployed in a relationship of trust.

A clearer example would be Chesley and Mary Yarn. Chesley carried on his business with his own ships in Mose Ambrose, and Mary had a complementary business with a ship, of which Chesley had made her the managing owner, in nearby English Harbour West. My visit with Mary in 1994 confirmed this arrangement.

Annie Northover Berkshire of Spencer's Cove, Placentia Bay, was appointed managing owner of their ship in 1959 and she sold 24 of her 40 shares to other family members. Her husband was a fisherman and perhaps he was recognizing his wife's schooling and business ability by placing her in charge of the ship. Eliza Pafford (Eddy) of North Harbour was married consecutively to two businessmen and was intimately involved in the operation of the business. For eleven years Laura Garland of Gaultois was manager of the Baddeck-built ship *Teresa G,* on which Thomas Garland Ltd. had taken a $75,000 mortgage. The mortgage was discharged after only 14 months. It is assumed that Thomas Garland was still living but only the firm is mentioned in the record. Mary Wilmot Sheaves of Channel shared an equal partnership with her husband.

Women who were traders, such as Ellen Roach of Branch, St. Mary's Bay, Ann J. Tuck of Fortune, Anne Farrell of St. Jacques, and possibly Gertrude Newman of Boyd's Cove, and Elizabeth Freake of Joe Batt's Arm, were certainly equal partners in terms of putting money into the ventures, sharing the risks and the profits, and very probably working on board ship during these coastal journeys. We saw that on the south coast women sometimes handled the money on board ship.

A whimsical story in the literature gives us a good example of a real partnership between husband and wife. Justus S. Wetmore and his wife of Clifton Village on the Kingston Peninsula of New Brunswick were partners in shipbuilding, farming, a sawmill and a granite quarry. Stanley T. Spicer writes in *Masters of Sail*: "that such was his role in the community and that of his wife that he was well known as 'He Boss' and she, 'She Boss.'"

d. **A woman is in charge; she is head of the operation.**

Mary Gosse of Spaniard's Bay was an excellent example of a woman in charge of an operation. She studied her six grown sons and appointed each to a specific task, and then operated the business until she was ready to pass it over. Alice Murphy of King's Cove, widow of Michael, operated the old MacBraire business for 30-plus years after her husband's death. Abigail Horwood of Harbour Le Cou also operated the family business for several years after her husband Todd died.

And, of course, the premier examples of independent leadership abilities were displayed in Ada Annie Petite of Mose Ambrose and Marie Smart Penny of Burgeo, and currently in Debbie Petite of English Harbour West.

I would conclude from these patterns and the cases cited that there seems to be little evidence that women were playing a passive role in ownership. I realize, of course, that the lack of precise documentation makes some of these conclusions tentative. But the weight of the material available seems to portray an active role taken by women.

To substantiate the case for dispensing with the near-myth of the passive women I can also cite the biographies I wrote in the summer of 2005 of fifteen women in the Bay of Exploits area, women who were active from the 1880s to the recent present. There was not a woman among them that I would characterize as passive. One, Mary Winsor, was a co-partner in a large sawmill operation in the Botwood area from 1889 to 1892. Another, Elizabeth Manuel, was the wife of the prominent Exploits Island merchant, Josiah Manuel. Josiah used to say: "I command the ship but my wife commands when I come off the ship." Michael Harris, in his book on the Crosbies, asserts that Elizabeth was the first one up every morning, ensuring that the men, maids and gardeners were at their chores. Another woman was Effie Cobb, owner and manager of the Transatlantic Inn in Botwood from the late 1920s to the early 1950s, when Pan-American Airways, British Airways, and American Export Airlines were using Botwood as a stopover between North America and Europe. Her hotel, restaurant and shop were located immediately adjacent to the headquarters of the three airlines and to the Canadian military base during World War II. Then there were the three Mi'kmaq women of Wigwam Point – Ellen, Fanny and Mary Paul – who were suppliers to the Winsor & Vallance Sawmill from the 1870s to the 1890s. There were the Evans women of Northern Arm – Hannah Eliza, Naomi and Selina – whose musically and dramatically talented daughters were constantly on call to the lodges and the A. N. D. Co. and for other social events in Botwood through the first years of the new company in the early 1900s. Another woman, Effie Taylor, born in Scotland and married to a World War I veteran from Botwood, became the first pensioned female from the A. N. D. Co. when she retired as stenographer in 1959.

In addition, there were many other women in the area who were midwives, nurses, postmistresses, bookkeepers, typists, stenographers, and so forth. Miss Lily Martin was transferred to Baie Verte in 1903 to manage the business firm of James Jackman, Esq.; she had been a clerk in Jackman's firm at Tilt Cove. Miss Harding of Pilley's Island was appointed stenographer to the very large Newlands Lumber Co. at Norris Arm in 1907, succeeding Walter Hopkins. And when James P. Howley of the Newfoundland Geological Survey completed his survey of the Botwood area, seven of the land speculators who bought sizable tracts of land were women, five of them married women. And elsewhere on the island, Miss Minnie Harding was appointed manager of the Samuel Harris, Esq., business at Grand Bank in 1907. None of these women, nor countless others of like mind, could be characterized as passive.

I also want to provide some details about the two women to whom this book is lovingly dedicated: my mother and my mother-in-law, both now deceased. My mother had only a grade-school education but she had worked as a domestic with a St. John's family and then as cook with the Herder family, of *Evening Telegram* fame, in the 1920s. She was worldly wise and used her experience in "the big city" to prepare me mentally and physically, with proper manners and social graces, when I left home in July 1948 to attend summer school in St. John's and again when I returned to Memorial University College in September 1948. My mother provided practically all the leadership in the family. From my father, being a man of few words, we learned only by example. My mother did not defer to my father, nor did she ever challenge him publicly. She complemented him in the sense that she supplied what he lacked. She was the real life and energy of the family. She set the tone with her ideas and her actions in emphasizing the need for a good education and in planning future careers for her sons (there were no daughters). My father was away from home a lot, working in the lumber woods and in a nearby town, and my mother planned and supervised the work for her sons: preparing and maintaining the gardens, making hay, ensuring that there was sufficient firewood, that the water barrel was always full, and that the animals were cared for and the barn and sheds kept clean. What self-fulfilment my mother experienced was through her involvement in church and community events and family visits. There was no time for competition between the sexes. Our home was a gathering place for women of the community, from young adults to young married women to older women. As young children we tried to eavesdrop on some of these conversations, which were conducted in whispered tones when they discussed family tragedies or grave community events.

My mother-in-law was from Twillingate and became a teacher at nearby Manuels and Crow Head. She married and moved to Botwood about 1927. As a young mother she enrolled in a correspondence course with the Chicago School of Nursing and completed the very detailed 53 lessons of 1937, which still exist today in a book that is fully three inches thick. On completion of the course she received a cap, pin and certificate. She thus became a resource person for many women in the community of Botwood, and there were women who would not allow the resident doctor to perform procedures on them unless my mother-in-law was present. She nursed her husband's two young brothers in her home until they both died and cared for her husband's father at her home until his death. She lost three young children and raised seven. She was among the first group of women in Botwood to be awarded a certificate and pin for Volunteer Nursing Service by the Canadian Red Cross after completing a requisite course of instruction and supervised practice which "fully qualified her to act as a nursing assistant without remuneration." She gave music lessons, learned shorthand in her later years, and was involved in leadership roles in church and lodge, at the local, provincial and national levels.

These were both proactive women. Compared with them, their husbands were passive.

Perhaps it is time to allow the near-myth of the passive woman to die, as the myth of the woman as bad luck at sea has died. The near-myth seems to have little basis in the reality portrayed here.

Women had to work within certain societal strictures which were characteristic of the times, but as we have seen repeatedly in this study they found ways of "pushing the envelope" and securing advances, one tiny step at a time.

The women described in this study deserve our admiration and praise. Women today stand on the shoulders of their sisters in the past.

APPENDIX A

Other Newfoundland Women in the Ship Registers

Many of the Newfoundland sole owners and joint owners are described in Chapters 2 and 3, but there are several others who are worthy of honourable mention. These are listed below with sufficient information to help determine how long women held on to ships and for what purposes these were used. The abbreviation M after the date indicates that the woman had arranged a mortgage on her ship.

Sole Owners

Owner, Place, Status/Occupation, Date, Name of Ship

Jane Parsons, Harbour Grace, Widow, 1837, ***Lavinia***

Jane bought this ship on November 13, 1837. Tobias Parsons served as master. No further transactions could be found.

Mary O'Neill, Fermeuse, Widow, 1844-5, ***Cousins***

Cornelius O'Neill was master of the ship. Mary sold the ship in 1845 to Dugald Markellar, merchant at St. John's.

Johanna Nowlan, St. John's, Widow, 1862-68, ***Dash***

When Michael Nowlan registered this ship in 1862, he transferred all 64 shares to Johanna, who was then a widow. Michael died on February 27, 1868, and his will, dated February 20, 1865, appointed

Johanna as executrix. She sold the ship on May 22, 1868, to Daniel Green, merchant of Harbour Grace.

Mary McCarthy, Carbonear, Widow, 1868-, *Alma*

Mary bought this ship on May 25, 1868, from Frederick Joseph Wyatt. The 44-ton vessel had been built in 1859 at Silly Cove. No further transactions could be found.

Sarah Andrews, Port de Grave, Widow, 1870-2, *Flower*

Sarah bought the ship in December 1870 and sold it in May 1872 to William and George Andrews, planters of Port de Grave, perhaps her sons.

Constance Garnier, Bay St. George, Trader, 1881-2, *Beatrice*

Ernest Garnier, "vice master" of the ship and possibly Constance's husband, is also listed as a trader.

Anne Farrell, St. Jacques, Trader, 1886-, *Carrie K.*

This vessel was built in Bridgewater, Nova Scotia, in 1882 and had been registered at Lunenberg in 1883. Thomas Farrell was master of the ship. Its name may originally have been the *Carrie McK.* Anne and Thomas Farrel were coastal traders along the south coast.

Kezia Foote, Grand Bank, Widow, 1889-, *Samuel J. Foote*

Kezia's husband had been Morgan Foote. Kezia died in 1910 at age 75.

Bridget Keough, St. John's, Widow, 1893, M, *Hope*

Bridget bought the ship from William Grant at Halifax on January 13, 1893, under Mortgage C. She sold it on June 13 the same year to James McAdam at Sydney.

Eliza Anne Tobin, Twillingate, Widow, 1893-6, *Erminie*
& St. John's, 1893-6, *Jubilee*
1893-7, *Dorothy*

Eliza Anne was the widow of Joseph Bennett Tobin, who was a merchant, major shipowner and member of the Grand Jury for Twillingate and who died February 14, 1893.

Laura Simmons, Harry's Harbour, Married, 1894-8, M, *Bonny Lass*

Laura was the first owner of this ship, buying it in 1894 from the builder, Thomas Burt. She may have had it built for her purposes. She took a mortgage on the ship in 1895 for the amount of $693.42 with interest at 6%; her mortgagee was George Knowling, merchant of St. John's. In 1898 Knowling transferred the mortgage to James Norris, planter of Three Arms, Notre Dame Bay. On November 14, 1898, Norris sold the ship to John McDonald of Salmonier. So it must have been Norris' way of collecting on a mortgage that could not be paid. It appears that Laura's business venture had failed. On November 21, 1902, John McDonald sold the ship to Anastassia McDonald, married woman of Salmonier, probably his wife.

Anna Vale McKenzie, St. John's, Married, 1894-1900, *Clara*

This 25-ton vessel was built in 1864 by the Cluetts of Fortune Bay who sold it to Anna in 1894. Anna had it rebuilt and enlarged to 27-tons in 1896 at Little Bay East by William Miles. On November 23, 1900, she sold the ship to Robert K. Bishop, merchant of St. John's. The ship lasted for a remarkable 36-plus years.

Margaret Catherine Fitzgerald, St. John's, Matron, 1896-, M, *Ada*

Margaret bought the ship on October 5, 1896, from John F. Apsey and Alfred G. Smith. It had been previously registered in 1878, so it was already 18 years old. Margaret took a mortgage for $500 with Robert G. Rendell, merchant. No further transactions could be found. The fate of the vessel is indecipherable in the record.

Eliza Strong, Old Perlican, Widow, 1897-8, *Mary Jane*

Eliza bought this 16-ton, nine-year-old ship and registered it on May 25, 1897. She sold it on November 10, 1898, to William C. Job, merchant of St. John's.

Emma Jane Chalk, Scissor's Cove, Married, 1898-, *Phoebe Jane*

Abraham Chalk, the builder of this 15-ton vessel, was probably Emma Jane's husband. No further transactions could be found.

Rachel Davis, St. John's?, 1899-, *Trio*

Rachel bought this 21-year-old ship after Walter Baine Grieve, merchant of St. John's, had discharged the mortgage of the previous owner.

Mary Murphy, St. John's?, Married, 1899-1900, M, *Kersage*

Mary had a mortgage with Charles W. J. Emerson for $100 and interest at 6% payable on November 15, 1900. She probably had bought the ship from George Graham Crosbie, merchant at Brigus. The ship was wrecked at Western Bay sometime in 1900. She may have been the Mary Murphy, aged 76, widow of Nicholas, who died in St. John's in 1916. If so, she left a husband and three sons.,

Elizabeth Ingraham, Ramea, Married, 1899-1902, *Nevada*

Elizabeth bought the 46-ton ship from her husband, John Ingraham, who had owned it for six years. The ship was owned later by Lena Young of St. Jacques.

Sarah Jane Johnson, Little Catalina, Married, 1899-04, M, *Lego*

James Johnson, the builder of this 16-ton vessel, was probably Sarah Jane's husband. Sarah Jane took a mortgage for $250 in 1900, one for $250 in 1901, one for $207.75 in 1902 and one for $200 in 1903; these were all at 10% interest. These were all discharged on June 2, 1904, and the next day she sold the ship to the mortgagee, Benjamin Garland, shoemaker of St. John's. It appears that business was not going well for the Johnsons.

Catherine Guy, Pilley's Island, Married, 1899-1927, *Esther*

Catherine bought this ship from her son Frederick after he had had it for one year and paid off his mortgage with W. C. Job, merchant, after only five months. The ship was lost at the Grey Islands in 1927, which means that Catherine may have held the ship for a remarkable 28 years. Frederick died in 1920.

Ann Andrews, St. John's? -1899, *Hyacinth*

Ann was the sixth owner of this ship; the purchase date cannot be determined. She sold it on June 10, 1899, to Michael Martin, trader of King's Cove. Martin sold it in 1902 to Emmie Wilcox and Henrietta Roberts, married women of Brigus. See Chapter 3.

Catherine Barron, Placentia, Widow, 1900, *Cygnet*

Frank Barron, mariner of Placentia, bought this ship on October 13, 1899. He died intestate on October 30, 1900. Letters of Administration were granted to Catherine by the Supreme Court of Newfoundland and she sold the vessel on November 25, 1900, to Walter Baine Grieve, merchant at St. John's.

Ann Barnes, Harbour Mille, Married, 1900-07, *Glencoe*

Ann's husband was John Barnes, the builder. Ann sold the ship in 1907 to Jeremiah Petite, merchant of Mose Ambrose.

Jessie Gushue, Brigus, Married, 1901-, *Jane*
1907-11, *Willis C.*

Jessie bought the *Jane* on December 15, 1901, but no further transactions could be found. She bought the 82-ton *Willis C.*, which had been built at La Have, Nova Scotia, on March 11, 1907, and sold it on June 19, 1911, to William C. Job, merchant of St. John's.

Julia Ann Johnson, Little Catalina, Married, 1902-, *D. B. J.*

Julia Ann's husband was probably Alexander Johnson, the builder of this 28-ton vessel. The fate of the ship is indecipherable in the record.

Mary J. Davis, Fox Harbour, Widow, 1904, *Bessie*

The 26-ton vessel had been built in Fox Harbour by George Lamb in 1900. Mary registered the ship on June 28, 1904, and sold it the same day to Reuben Pike, mariner of St. John's.

Mary Horwood, Carbonear, Married, 1905-, *Maggie Barbara*

Samuel P. Horwood, planter of Carbonear, bought the 22-ton ship on April 8, 1905, and sold it on April 11, 1905, to Mary Horwood. No further transactions could be found.

Charlotte A. Haliburton, Bonne Bay, Married, 1905-14, *Marie Ann*

This 30-ton ship had been built in Quebec in 1890. Louisa A. Whiteley had owned it from 1903 to 1905. Charlotte's husband was probably John P. Haliburton. The vessel was broken up at Bonne Bay in January 1914.

Lena Young, St. Jacques, Married, 1905-9, *Nevada*

This ship had been owned previously by Elizabeth Ingraham of Ramea. The Youngs had been a merchant family at St. Jacques since the 1880s.

Maria Tuck, Fortune, Widow, 1905-17, *Argo*

Maria was the widow of William Tuck who built the Argo in 1895 and died on December 24, 1905, and she was also executrix of his will. She sold the ship on May 1, 1917, to Walter John and Alfred

Tuck, mariners of Fortune, and probably her sons. Maria died on April 24, 1933, at age 84.

Jemima Emine Parsons, Freshwater, Married, 1905-7, *Orient*

Jemima bought the ship from Francis J. Parsons, planter of Freshwater, Bay de Verde, on May 25, 1905. The ship was lost at Wesleyville in September 1907 or 1917; the date is indistinct in the record.

Margaret Ryan, Catalina, Married, 1906-17, *Success*

Margaret's husband was P. F. Ryan and she was the youngest daughter of John and Johanna Murphy. Margaret died in November 1917 and left a husband, two sisters and three brothers. The 17-ton vessel was broken up at Catalina about 1919.

Frances Miles, Herring Neck, Planter, 1906-12, *Pratincole*

This 64-ton ship was built at Exploits Harbour by Robert A. Sceviour for Thomas A. Winsor, merchant of Exploits. Winsor sold it on December 24 to Frances Miles, planter. This is the only time "Planter" was used in the ship registers for a woman. Both times the name "Frances" appears in the record, it is clearly spelled with an "e." Henry, Esau and Stephen Myles of Herring Neck were planters from the 1830s to the 1880s. Many of the early women were planters in all but name only, e.g., Mary Ludevig of Ferryland and the women of Spaniard's Bay. Frances sold the ship in June 1912 to William Ashbourne, merchant at Twillingate.

Susannah Davis, Harbour Grace, Married, 1906-10, *Pleasant Time*

Susannah bought this ship from John T. McRae, merchant of Harbour Grace, on February 14, 1906. The ship was lost on the Labrador on September 20, 1910.

Teresa Power, St. Mary's, Widow, 1906-11, *Bridie M. Power*

Teresa inherited the ship when her husband, James Power, died without a will on August 7, 1906. She held on to the vessel for five years and sold it on May 8, 1911.

Carrie Carter, Summerside, Widow, 1907-, *Erita & Ena*

Herbert D. Carter built the *Erita & Ena* in 1905. Carrie was the owner when she registered the 33-ton vessel on July 12, 1907. No further transactions could be found.

Jessie A. Burdock, Belleoram, Married, 1910-11, *Gertrude*

Jessie bought the *Gertrude* at St. Pierre et Miquelon on December 17, 1910; the ship had been built at Essex, Massachusetts. She sold it on June 26, 1911, to Robert Moulton, merchant at Burgeo.

Ellen Snow, Bay Roberts, Married, 1913, *Darryl*

Ellen bought the ship on May 28, 1913, from William H. Peters, merchant of St. John's, and sold the ship on June 7, 1913, to Philip Templeman, merchant of Bonavista, owning the ship for only 10 days.

Ursula Petipas, Curling, Married, 1914-, *Kathleen Alice*

Ursula bought the 13-ton vessel the same year it was built at Curling; it had a vertical reciprocating internal combustion engine. The ship was broken up at Flower's Cove in 1947.

Martha Carter, Channel, Widow, 1915-16, *Good Ship Jubilee*

It is difficult to figure out the history of the *Good Ship Jubilee*. William Carter, master mariner of Channel, had the 59-ton vessel built for him at Channel by George Scott in 1897. Sometime after that he appears to have sold it to Philip T. Clement and held a mortgage on the ship. Martha may, in fact, have been the mortgagee. When William Carter died, perhaps in 1914 or early 1915, Clement sold the ship to Martha on May 22, 1915, perhaps in settlement of the mortgage. On that date Martha is referred to as the "Administratrix for the estate of William Carter, late of Channel, mariner, deceased." On June 21, 1916, Martha sold the ship to Theodore Keeping, mariner of Channel.

Catherine Reddy, Marystown, Widow, 1915-18, *Woodrow Wilson*

When Hugh Reddy, merchant of Marystown, died on April 6, 1915, his will of May 9, 1912, appointed Catherine as executrix. Catherine was a nurse from England and had a long career in nursing before she married Hugh. She sold the ship on September 21, 1918.

Susannah R. Baggs, Curling, Widow, 1918-19, *Brunswick*

James H. Baggs, General Merchant at Curling, bought this ship in 1917 and died intestate on February 24, 1918. Letters of Administration were granted to Susannah Baggs and William Roberts. Roberts was a physician at St. John's. They sold the ship on March 28, 1919, to the St. George's Steamship Co. Ltd.

Catherine Wade, Conception Harbour, Married, 1919-44, ***James P. Wade***

Catherine's husband was probably James P. Wade, merchant, from whom she bought the ship on May 29, 1919. If Catherine still owned the ship when it was broken up on the Labrador in 1944, she would have been the owner for 25 years!

Mary C. Kennedy, Carbonear, Married, 1921-, ***Coronation***

Mary bought the 40-ton vessel on June 14, 1921, under Mortgage A from A. E. Hickman Co. Ltd. No further transactions were found.

Eleanor Roberts, Twillingate, Married, 1921-22, ***Latona***
1923-27, ***St. Helena***

Eleanor bought the *Latona* on August 18, 1921, from John T. Moulton, merchant of Burgeo, and sold it on March 22, 1922, to Edward Roberts, merchant of St. John's. She bought the 36-ton *St. Helena* on May 25, 1923, and sold it on January 15, 1927, to Henry William Rideout, contractor of Twillingate.

Agnes Davis, Fox Harbour, Widow, 1922, ***Clematis***

Agnes bought the 27-ton vessel on March 24, 1922, from William C. Job, merchant of St. John's, and sold it on April 24, 1922, to a fisherman at Petit Forte, having had it for one month only.

Jessie Beatrice Bragg, St. John's, Married, 1922-3, ***Earl Grey***

Jessie bought the ship on January 3, 1922, from Marmaduke G. Winter, merchant of St. John's, and he held a mortgage for $5,000, which was discharged on June 26, 1923. Jessie sold the ship the very next day to a mariner from Coachman's Cove.

Anne A. Barry, Curling, Married, 1922-27, ***Philomena***

David Barry of Bay of Islands built the 50-ton *Philomena* and installed a compound condensing engine with a Scotch boiler made by the Portland Engine & Co. Ltd. It was registered in Anne Barry's name. The vessel was totally lost at Caribou Brook, Cape St. George, August 26, 1927.

Margaret Davis, Argentia, Married, 1923-, ***P. & W. Davis***

Margaret was the first owner of the 28-ton ship. A new certificate was issued for the ship on May 18, 1923, because the original one had been lost. An attempt was made in July 1952 to trace the vessel and on August 8, 1955, it was restored to the registry. The certificate was canceled on March 28, 1967, due to a report from a Customs and Excise officer that the "vessel drifted ashore at Harbour Breton during a storm a few years ago and considered unfit for further service." It is impossible to tell from the record how long Margaret held the ship, but it seems remarkable that the ship may have lasted for almost 40 years.

Rebecca Chetwynd, Port aux Basques, Widow, 1924, *Luxor*

John P. Chetwynd, merchant at Port aux Basques, had owned this ship since 1917. When he died on February 5, 1924, his will of November 6, 1912, appointed Douglas J. Chetwynd and Rebecca Chetwynd as executors. Douglas was a clerk, perhaps the son of John and Rebecca. After probate was granted, they sold the ship on November 8, 1924, to Francis Pike, merchant at Port aux Basques.

Mary Ann Bennett, Port au Bras, Burin, Widow, 1924-25, *James U. Thomas*
1928, *Harvey & Maud*

Samuel Bennett, merchant of Port au Bras, bought the *James U. Thomas* in 1921. He died intestate on October 28, 1924. He may have been the "Old Sam" Bennett who was drowned in the St. John's Narrows when he was knocked overboard by the swinging main-boom. Letters of Administration were granted to Mary Ann and she sold the vessel on February 18, 1925, to A. E. Hickman & Co., merchants of St. John's.

George Bennett, mariner of Port au Bras, bought the *Harvey & Maud* (it had previously been named the *Lilian Walsh*) on June 15, 1922. On March 15, 1928, he sold the vessel to Mary Ann Bennett and just one week later, on March 22, Mary Ann sold it to John Warren, merchant at Gaultois.

Mary E. Bailey, St. John's, Married, 1924-29, *George Wheatley*

Mary became the fifth owner of this 65-ton ship which had been built by Jabez Manuel of Exploits in 1906. She bought it on April 8, 1924, from John T. Currie of Britannia and sold it on June 25, 1929, to A. H. Murray & Co. of St. John's.

Minnie Ellis, St. John's, Spinster, 1926, *Admiral Drake*

Minnie Ellis and Samuel Ryan, accountant, were appointed executors for the estate of William J. Ellis when he died on April 19, 1926. The ship, which he had bought in August 1925, was only a small part of his estate which was valued at $140,000. Ellis was born in 1857, had apprenticed as a stonemason with his father, and at the time of his death had an unmarried sister, Minnie, and three nieces living with him in St. John's, and a widowed sister, Mrs. McManus, living with him in a house he owned in Brooklyn, New York. He had been spending the winter in California and died on his way back to St. John's. Minnie Ellis and Samuel Ryan, as executors, sold the *Admiral Drake* on August 9, 1926, to A. S. Rendell & Co., merchants of St. John's.

Hannah Boyce, Jersey Harbour, Widow, 1926-7, *Jean Dorothy*

Daniel Boyce, merchant and also Hannah's husband, died in 1926. Hannah sold the *Jean Dorothy* to D. Boyce & Sons Ltd., undoubtedly her sons, on May 15, 1927. See Hannah also in the "Joint Owner" list.

Sarah Ann Mitchell, Marystown, Married, 1926-28, *Mary Mitchell*

After having had the vessel for nine years, Edward Mitchell, fisherman of Marystown, sold it to Sarah Ann. After holding it for two years she sold the ship to a fisherman at Trepassy.

Laura Greening, St. John's, Spinster, 1928, *Lizzie E. B.*
1928, *Lewisport*
1928, *Vigilant*

Joseph Morris Butler, storekeeper of St. John's, owned all three ships and died on January 11, 1928. His will, dated January 8, 1928, appointed Laura Greening as executrix. Laura acted with despatch and sold the *Lizzie E. B.* on April 4, 1928, to William Hiscock, fisherman of Flat Islands, Bonavista Bay, the *Lewisport* on April 24, 1928, to A. H. Murray & Co. Ltd., and the *Vigilant* on June 27, 1928, to Gavrock Brown of Fair Islands.

Emma Parsons, Port Union, Widow, 1929, *Humber Deal*

Ernest Parsons, fisherman of Port Union, bought this ship in 1924 and died intestate on April 24, 1929. Letters of Administration were granted to Emma Parsons on June 11, 1929, and she sold the ship one day later on June 12, 1929, to Joseph Parsons, planter of Newtown.

Zealah Snow, Bay Roberts, Widow, 1929-32, *Reliance*

Zealah inherited this ship when her husband, Edward Snow, died on January 6, 1929. The 48-ton vessel had been built in 1903 at Northern Arm by Robert Evans, my great uncle. Zealah sold the ship on March 22, 1932, to a fisherman at Fortune. This vessel was lost "somewhere on the Newfoundland coast" in 1936, thus lasting for a remarkable 33 years.

Alice M. Currie, Britannia, Widow, 1930, *Clarence B.*

Alice inherited this ship when her husband, John T. Currie, died in 1930. She sold the ship after only 26 days, but she held on to her husband's other four ships, one of which was named the *Alice*, and continued his business for another 10 years. Currie was "the big businessman for Smith Sound"; he did coasting work as well and fishing on the Labrador. Alice was a Laite from Petley; her parents were Deborah and Uriah. Reuben Vardy became Alice's bookkeeper after John's death. Alice's brothers – Charles, Willis and Absalom – skippered her ships. Another brother, Rev. Hugh Laite, was a Japanese prisoner-of-war for 44 months. Alice had five sisters, including two sets of twin sisters. She and John had one son, Ralph. When Alice died in 1940 the Currie business folded. Alice Jerrett of Brigus owned the *Clarence B.* for a short period in 1935.

Mary Ellen Carter, St. John's, Widow, 1930-38, *Frank H. Adams*

Albert Haynes, master mariner of Catalina, was the owner of this ship. Thomas H. Carter, ship's broker of St. John's, was the mortgagee for Albert for $9,500 with interest at 6% from March 11, 1920. When Thomas Carter died on June 20, 1930, his will appointed Mary Ellen as executrix and she took over the mortgage. She discharged the mortgage after eight years on April 16, 1938.

Hannah Parsons, Clattice Harbour, Married, 1931-, M, *Catherine P. Follett*

The builder, Peter Follett of Clattice Harbour, sold the 12-ton vessel to Hannah on March 12, 1931. She took a mortgage with James Baird Ltd., merchant of St. John's. No other transactions on this vessel could be found.

Sarah Pittman, Merasheen, Married, 1931-33, *Jenetta Pittman*

The 19-ton vessel was built in Harbour Buffett by Benjamin Foote in 1888. Patrick Pittman, fisherman of Merasheen, rebuilt the ship in 1931 and sold it to Sarah on December 19, 1931. Sarah died without a will on October 14, 1933, and Letters of Administration were granted to Patrick. He sold it in 1938 to Thomas Stanley Keats, merchant at Argentia. The vessel was then 50 years old!

Araminta Frances Collishaw, St. John's, Widow, 1931, *Julia R. Blackwood*
1931, *Swile*
1931, *Young Hood*
1931-49, *Stella Hynes*
1931, *L. M. May*
1931, *F. P. Union*
1931, *Gertrude*

Araminta's husband, Ellison Collishaw, a merchant in both Halifax and St. John's, died on June 22, 1931. His will of October 21, 1930, appointed Araminta and the Royal Trust Co. as executors. They sold all ships within one year to the Fisherman's Protective Union Trading Co. at Port Union, except the *Stella Hynes* which Araminta held on to for 18 more years and then sold to the F. P. U.

Blanche Louisa Newman, Harbour Le Cou, Married, 1932-34, *Agnes R.*

Blanche Louise bought the ship from Thomas Rose, merchant at Harbour Le Cou, on May 6, 1932. She sold it on February 10, 1934, to Simeon Butt and subsequently it became the property of Violet Blanche Butt, also of Harbour Le Cou.

Emma Frances Samson, Flat Islands, B.B., Widow, 1933-34, *Lloyd Elsworth*

Lewis Samson, fisherman of Flat Islands, bought this ship in November 1922 and died without a will on May 23, 1933. Letters of Administration were granted to Emma Frances and she sold the ship on November 5, 1934, to Samuel M. J. Barbour and Leslie M. Barbour, fishermen of Newtown. Whether the ship was used for the two fishing seasons after Lewis' death is not known.

Mary J. Parrott, St. Bernard's, Married, 1933-38, M, *Mary R. Parsons*

Mary's mortgage with Sir Marmaduke Winter was discharged through the Eastern Trust Co. in 1938 (Sir Marmaduke died in 1936) and she continued to hold the ship even after 1938, but it cannot be determined for how long.

May B. Pike, Port aux Basques, Married, 1934-36, *Hi Sea Bird*

E. Pike Ltd. (Emanuel?) had a store and ship business, and Mary was very much involved in running the business. The 9-ton motor screw vessel was lost at Port aux Basques in 1936.

Margaret Brenton, Burin, Married, 1934-43, *Margaret Brenton*

Margaret was a World War I bride from Scotland and her husband John was a sea-going captain. Margaret named the 23-ton ship after herself when it was built by John W. Kendall, lumber merchant at Morrisville in 1934. Margaret had the ship altered and an auxiliary motor screw installed in 1940, giving the ship a four knots speed. In 1943 she sold the ship to Thos. Garland Ltd. of Gaultois.

Emily Young, Twillingate, Married, 1934-43, *Edward and Sons*

Emily bought the 13-ton sailing vessel when it was built by Elisha Wheeler at Virgin Arm. She sold the ship on April 14, 1943, to two merchant brothers at Twillingate.

Alice Jerrett, Brigus, Widow, 1935, *Clarence B.*

Alice M. Currie of Britannia had owned this ship briefly in 1930. Alice Jerrett inherited the ship when her husband, Charles A. Jerrett, merchant of Brigus, died on April 7, 1935. Alice and her fellow executor, Ernest Victor Jerrett, merchant of Brigus, sold the ship on June 13, 1935, to George Churchill Jerrett, merchant of Brigus, thus keeping the ship in the family.

Marion Owen Porter, Flower's Cove, Married, 1935-, *White Gull*

This small 10-ton vessel was built by George Dalton at Little Catalina and fitted with an auxiliary motor screw; the vessel was lost at Woods Island, Bay of Islands, date unknown

Ella Perpetua Lewis, St. John's, Married, 1935-, *Theresa M. Gray*

This ship was built at Port Clyde, Nova Scotia, in 1902 and transferred to the St. John's Registry in 1911. On September 6, 1935, when the ship was already 33 years old, Ella bought it from Baine Johnston & Co. Ltd. No further transactions could be found.

Emily Clarke, Harbour Le Cou, Married, 1935-48, *Blanche Clarke*

Emily Billard was married to David Clarke, fisherman. Emily bought the 19-ton ship in 1935 and had a motor installed and the ship registered *de novo* in 1936. She sold the ship in 1948 and it was transferred to Sydney, Nova Scotia.

Caroline Hannam, Rose Blanche, Married, 1936-53, *Wilehema*

Caroline Hatcher was married to William Hannam. Caroline bought the 15-ton ship, which was built by Josiah Farrell of North Bay, LaPoile, in 1935. In 1936 she had a motor installed and registered the ship *de novo*. After 18 years of ownership, in 1953, she sold the ship to Charles Hannam, fisherman, and Mary Elizabeth Hannam, married woman of Rose Blanche. See Mary Elizabeth Hannam in the "Joint Owner" list.

Elsie May Blackwood, Safe Harbour, Married, 1936-37, *A. H. Whitman* 1940-47, *Tessie Aubery*

On June 10, 1937, the *A. H. Whitman* ran on a rock off Flagstaff Run 11 miles from Hopedale, Labrador. She was pulled off with the aid of motor boats and towed near the shore but was leaking so badly that she was abandoned and subsequently sold by the Wreck Commission as a constructive total loss. She was eventually restored to the registry under a new owner.

Elsie May's second vessel, the *Tessie Aubery* was totally lost at Split Point entrance to Baccalieu Tickle on July 19, 1947. The vessel had been fitted with an auxiliary motor screw. See Elsie May Blackwood in the "Joint Owner" list for two more ships.

Fanny Isabel Ryan Fiander, Trinity, Married, 1937-38, M, *Gertrude Jean*

This was a crude oil screw previously registered in 1932 in Bridgetown, Barbados. It had been built in 1930 by Marie Amy Deveau at the Meteghan Shipbuilding Co. in Nova Scotia. Fanny bought the vessel in 1937 and took a mortgage for $2,000 at 3% interest with John Ryan, police officer retired, possibly her brother. Her mortgage was discharged on March 31, 1938, and on April 14 she sold the vessel to Edmund Vardy, mariner of Hickman's Harbour, and Edmund's mortgagee was Lillian Vardy at the amount of $4,000.

Ellen Kenny, Fermeuse, Widow, 1937-49, *M. A. Kenny*

Walter Kenny (of Andrew), fisherman of Fermeuse, had owned this ship since 1930. In 1936 he paid off his 1932 mortgage with Walter S. Monroe, merchant of St. John's, and died on February 7, 1937. His will of January 26, 1937, appointed Ellen and John Joseph Kenny, fisherman, executors. Probate was granted in August 1939 but they must have continued in the fishery for a total of 12 years and sold the ship in April 1949 to two fishermen of Change Islands.

Mary Burke Norris, St. John's, Widow, 1938-39, *Granite*

This 120-ton vessel was built at Lunenberg, Nova Scotia, in 1913. Its foreign name was *Poseidon*. The fourth owner, Ambrose F. Norris, master mariner of St. John's, died intestate on July 5, 1938. Letters of Administration were granted to Mary and she sold the vessel on April 5, 1939, to John Hann Blackmore, mariner of Port Union.

Mary Canning, Presque, Widow, 1938-42, *A. B. Flynn*

Mary inherited this ship through her husband Peter Canning's will when he died on January 5, 1938. Peter had been a merchant at Presque. It is likely that Mary carried on the business; she sold the ship to another merchant on August 17, 1942.

Mary Maloney, Sweet Bay, Widow, 1939-44, *M. G. Butt*

Mary inherited the 36-ton ship when Stephen Maloney, fisherman of Sweet Bay, died without a will on November 6, 1939. She must have carried on the fishery for more than four years and sold the ship on May 30, 1944, to Earle Sons & Co. Ltd. at Fogo.

Susie Rebekah Peckford, Change Islands, Spinster, 1941-42, *Lester N. Peckford*

Charles William Peckford, fisherman, bought this 33-ton ship in June 1938 and when he died on May 15, 1941, he left his entire estate, including the ship, to his niece Susie who was his housekeeper and care-giver in his last years. On May 2, 1942, Susie sold the ship to Sidney Macdonald Peckford, fisherman. Susie later married Max Scammel and they had a son Eddie.

May Fowlow, Trinity East, Spinster, 1942-43, *Millicent & Joan Fowlow*

This 71-ton ship had been built in 1906 at Lawrenceton by George Manuel and rebuilt in 1930 by Edmund Frampton at Smith Sound, Trinity Bay. It had originally been named the *St. Clair*. Albert Fowlow, mariner of Trinity, registered the vessel *de novo* in June 1930. He died without a will on October 2, 1942. Letters of Administration were granted to May Fowlow, undoubtedly Albert's sister, and she sold the ship on January 14, 1943.

Cecilia Flynn, Bar Haven, Married, 1942-58, *Mary & Stanislaus*

This 15-ton schooner had an auxiliary motor screw. The vessel was condemned in June 1958 as being unfit for further service.

Lucy Embree Button, New Melbourne, Widow, 1941-43, *Grilse*

Lucy bought the 32-year-old vessel in 1941 from a fisherman in Placentia Bay and sold it on May 11, 1943, to a fisherman at Change Islands. Another long-lasting ship.

Jessie Bragg, Port Union, Widow, 1943-64, *Marie Yvonne*

Jessie was probably the widow of John Bragg, master mariner, who died in 1939, and the sister-in-law of Walter Bragg, mariner and executor of John's estate, from whom she bought the vessel. The *Marie Yvonne* had been built in 1931 by Marie Amy Deveau of the Meteghan Shipbuilding Co. in Nova Scotia. Jessie sold the ship on May 19, 1964, having had it for a remarkable 21 years. She did not have a mortgage on the ship; the mariner and salesman of St. John's who bought it had to take a mortgage.

Stella Louise Noseworthy, Fortune, Married, 1943-44, *Clara Murray*
1944-46, *Helen G. McLean*

The *Clara Murray* had been built at Fox Harbour, Placentia Bay, in 1907 by Frank Murray and rebuilt in 1927; Stella bought it on April 24, 1943, and sold it on January 31, 1944. The *Helen G. McLean* had been built in 1913 at Shelburne, Nova Scotia. Stella bought it on January 17, 1944, and sold it on April 20, 1946. Stella had a penchant for old ships.

Lillian Warr, Pilley's Island, Widow, 1944, *Nina L. V. Davis*

Harold George Warr, merchant of Pilley's Island, died on January 6, 1944. His will of July 18, 1930, appointed Lillian Warr as executrix. She renounced probate of the will, and Letters of Administration were granted to Reginald Willie Warr, merchant of Springdale, probably her son. This was the only case found in the ship registers of someone renouncing probate of a will.

Muriel Kendel, Ramea, Widow, 1944-45, *Annie K.*

John Kendel, fisherman, was the builder of this 11-ton vessel in 1936. He sold it in 1937 to John Kendel, Jr., who died without a will on April 24, 1942. Letters were granted to Muriel on April 24, 1944. She sold the vessel on June 6, 1945, to John Penny & Sons Ltd. of Ramea. Presumably the vessel was being used in the fishery from 1942 to 1945.

Pearl Goodyear, Carmanville, Widow, 1945, *Radio City*

The record on this ship is very unclear. Stewart Hindoo Goodyear, mariner of Carmanville, bought the 56-ton vessel from Job Bros. & Co. of St. John's in November 1940. Job Bros. & Co. held a mortgage for the new owner and this was discharged in December 1944. Stewart Goodyear may have died but there is no reference to a death or a will. Instead Pearl Goodyear is listed as becoming the fourth owner of the ship on May 26, 1945, "under powers vested in her by Order of the Supreme Court of Newfoundland." She sold the vessel the same day to Paul Prestwood Small, merchant of Bridgeport, Notre Dame Bay. Jane Bartlett of Rattling Brook had owned this ship from 1935 to 1936.

Catherine Jane Symes, Rencontre West, Married, 1945-46, *Merle and Agatha*

This ship had been built at White Head, Nova Scotia, in 1929 and bought by Arthur M. Symes of Rencontre West in 1944. On March 10, 1945, he sold it to Catherine Jane, and she sold it on January 4, 1946, to Job Bros. & Co. Ltd. at St. John's.

Fredrica Hunt, Fair Island, Married, 1946-47, *Maggie Stone*

The *Maggie Stone had* been built in Lunenberg, Nova Scotia, in 1907 and rebuilt in 1925 by Emanuel Stone at Monroe, Trinity Bay. It was named for Emanuel's wife, Margaret. Fredrica Hunt bought the ship from Vincent James Guy, merchant at Catalina, on May 23, 1946, and sold it on May 31, 1947, to S. W. Mifflin Ltd., merchant at Catalina. Margaret Stone had owned 11 shares in this ship from 1907 to 1918.

Annie Sheaves, Port aux Basques, Married, 1946-47, *Bessie Louise*

Annie bought this ship on March 1, 1946, from S. W. Mifflin, merchant of Catalina, and sold it on May 29, 1947, to Harvey Taker of Grand Entry, Magdalen Islands. George Sheaves and Annie had a small business in Port aux Basques. They later moved to North Sydney.

Helen Rosina Amelia Monroe, St. John's, Married, 1947-, *Zipper*

The *Zipper* had been built in 1943 at South Rondout, New York, and was originally named the *Chauvenet*. Within 23 days after she had bought it on October 7, 1947, Helen Rosina sold the ship to Zipper Ltd., St. John's. This may have been a business firm under the Monroe firm's control. Arthur Harvey Monroe bought the ship on May 12, 1947, and sold it to Helen five months later.

Amelia Foote Patten, Grand Bank, Married, 1947-55, ***Miss Glenburnie***

Amelia was the daughter of Robert and Sarah Dunford. Her husband was Howard Patten, who was a prominent businessman in Grand Bank. One month after purchasing the *Miss Glenburnie*, Amelia had alterations made to the ship and an engine installed. Her ship was cut down and became a total loss near Egg Island Buoy on the Nova Scotia coast on May 24, 1955. Amelia died in 1972 at age 72.

Lucy Ella Piercey, Grand Bank, Widow, 1948, ***Ariel***
1948-55, ***Harold Guy***

Samuel Piercey, merchant of Grand Bank, died on January 1, 1948. Lucy Ella and Elic Forward Piercey, merchant, were appointed executors through Samuel's will dated January 29, 1937. They sold the *Ariel* to Samuel Piercey Evans, student. Two months later the vessel caught fire near Bay Bulls and was abandoned and then towed to St. John's where it became a constructive total loss. The *Ariel* had been built in 1921 in Sureines, France.

Lucy Ella and Elic Forward Piercey did not apply for probate concerning the *Harold Guy* until January 22, 1955; they then sold the ship three months later. The ship must have been in use during that seven-year period.

Hilda Mary Tibbo, Harbour Breton, Widow, 1950-51, ***Myra Kathleen***

William Tibbo, Sr., mariner, died intestate on March 30, 1950, and Letters of Administration were granted to Hilda Mary on October 5, 1950. She sold the ship exactly one year later, on October 5, 1951. Hilda Mary was William's second wife and was only 14 years old when they married. For years William operated a passenger boat named the *Hilda M.* between Harbour Breton and Grand Bank. He also owned a schooner named the *M. V. Minnie Pearl*, and Frances Jensen's husband, Hermon, operated a store on this schooner. Frances ("Fan") accompanied her husband on these coasting ventures and they brought home "boxes of money" at the end of each venture. Hilda Mary Tibbo later moved to Grand Bank and married George Crant.

Edith Blanche Bishop, Wesleyville & St. John's, Widow, 1950-52, *Lloyd Jack*
Eldon Bishop, fisherman of Wesleyville, died without a will on July 16, 1950. Letters of Administration were granted to Edith on May 2, 1952. At that time she was residing in St. John's. The vessel was lost at Shambler's Cove, Bonavista Bay, July 14, 1952; it was being operated at the time by W. J. Pittman of Pilley's Island.

Audrey Day, Gaultois, Married, 1951-, *Willie & Ross*
Audrey bought this ship on August 23, 1951, from Archibald Billard, fisherman of Grand Bruit. The registry closed on this vessel September 10, 1968. It had been lost at sea some years previously.

Evelyn Ruth Hann, Wesleyville, Married, 1952-53, *Redwing Chief*
Jesse Hann, insurance underwriter of Wesleyville, died without a will on January 5, 1951. Letters of Administration were granted to Lloyd George Hann, manager. Evelyn Ruth Hann bought the vessel on March 31, 1952, and appointed Lloyd George Hann as "the person to whom the management of the vessel is entrusted by and on behalf of the owner." The ship burnt and became a total loss at Hermit Cove, Bonavista Bay, on December 2, 1953. Evelyn Ruth informed the registrar's office in writing.

Frances Mary Grandy, Garnish, Married, 1952-55, *Earl John Miller*
Frances bought the ship on December 11, 1952, from Thomas Grandy, mariner of Garnish. She appointed Thomas as "the person to whom the management of the vessel is entrusted by or on behalf of the owner." In October 1955 she sold the ship to John Rose, mariner of North Sydney. In August 1958 the ship was bought by Marion Emerson White, married woman of Murray Harbour, Prince Edward Island.

Bridget Power, Little Bay, Marystown, Married, 1953-55, *Lucy & Melinda*
Bridget bought the ship on January 16, 1953, from three fishermen of Little Bay by the name of Power. On January 28, 1953, she appointed John Patrick Power, one of the fishermen from whom she bought it, as "the person to whom the management of the vessel has been entrusted by or on behalf of the owner." The vessel went ashore at the entrance to St. Pierre on November 5, 1955, and became a constructive loss.

Irene May Saunders, Cobbs Arm, Married, 1959-60, *Fortunegull*

Irene May bought the ship, probably in 1959, from Oliver Saunders, fisherman of Cobbs Arm. On December 8, 1959, "advice under the hand of Irene May Saunders" was received at the registrar's office, appointing Oliver "ship's Husband." The vessel ran ashore on Shallop Rock off Noggin Cove, Notre Dame Bay, and became a total loss on November 7, 1960. "Advice received from owner" by the registrar.

Gladys Irene Newman, Petites, Married, 1960-63, *Ralph & Bob*

Robert Allan Newman, merchant of Petites, sold the ship to Gladys Irene on December 17, 1960. She appointed Robert as managing owner of the ship. The ship was sold to United States citizens, but probably by a later owner; the registry closed in 1968.

Mary Ellen Grace Moores, St. John's, Married, 1960-72, *Tedcap*

The 2-ton yacht had been built in 1945 in Upper La Have, Nova Scotia, and had formerly been called the *Kayfra*. Mary Moores bought the yacht on April 20, 1960, and informed the registrar's office on April 28 that she was appointing herself as managing owner. She sold the ship on June 9, 1972.

Gladys Marion Wiscombe, Creston, Widow, 1961-65, *J. W. Wiscombe*

Gladys bought the 95-ton vessel from the trustee of the bankrupt sale of Hollett Sons & Co. of Burin on September 18, 1961. She changed the name of the vessel from *Trinity North* to *J. W. Wiscombe* in honour of her late husband, merchant at Creston. She took a mortgage on the ship. The ship was totally lost in January 1965 when it ran aground at Point May.

Selina Frances Vallis, Coombs Cove, Married, 1961-65, *John C. Musseau*

Selina bought the 15-ton vessel on November 1, 1961, and on December 12 appointed Walter Chesley Vallis of Coombs Cove as managing owner, i.e., "the person to whom the management of the vessel is entrusted." The record states, as in all cases of this kind: "Advice under the hand of Selina Frances Vallis received 12 December 1961."

Evelyn Whiffen, Bonavista, Widow, 1963-4, *Anita Joyce*

Andrew Whiffen, fisherman of Bonavista, had owned the ship since 1955; he died without a will on August 3, 1963. Letters of

Administration were granted to Evelyn Whiffen, widow, on April 9, 1964. She sold the ship on July 3, 1964.

Rita McDonald, Salmonier, Married, 1964-65,
Halford Robert & Lawrence

John Joseph Hickey of O'Donnells, St. Mary's Bay, bought the ship on August 4, 1964, and the next day took a mortgage with Rita McDonald. She discharged Mortgage B on December 23, 1965.

Joint Owners

Owner, Place, Status/Occupation, Date, Name of Ship

Ellen Murray, Carbonear, Widow, 1828-, *Hope*

Ellen's husband, John Murray, planter of Carbonear, was a joint owner of the *Hope* with Robert Pack, John Goss of Poole and William Fryer of Wimborne in the County of Dorset, merchants. John Murray probably was the master of the ship, which had been built at Belle Isle in 1805 and "enlarged under the hand of William Rockwood builder dated 15 March 1827." John Murray died on June 7, 1828; he had bequeathed by his will of June 28, 1827, his 32 shares in the *Hope* to his widow Ellen. The record ends with a note which leads one to believe that Ellen remained a majority partner with the three merchants.

Olivia Emma Winsor, Wesleyville, Widow, 1908-21, *Tasso*

Olivia's partners at the time of purchase on January 14, 1908, were Jesse Winsor and William C. Winsor, though she was the first-named owner. She sold her shares and rights to them after 13 years, on October 5, 1921.

Catherine Saunders, Aquaforte, Widow, 1916, *David M. Hilton*

This ship had been built in Essex, Massachusetts, in 1888. Richard Saunders owned 16 shares in this ship. He died without a will on April 8, 1916. Letters of Administration were granted to Catherine on August 3, and she sold the 16 shares just 18 days later.

Alice I. Butler, St. John's, Spinster, 1918-26, *Admiral Drake*

Alice inherited two shares when William F. Butler, architect of St. John's, died without a will on February 24, 1918. She was still holding the shares in 1926. Minnie Ellis, spinster of St. John's, became an executor for the estate of William J. Ellis and was a joint owner of this ship for four months. See Minnie in the "Sole Owner" list above.

Agnes H. Fitzgerald, St. John's, Stenographer, 1919-28, *Emily E. Selig*

Agnes bought 2 shares in this vessel on May 29, 1919, from A. S. Rendell & Co. Ltd. The vessel was totally lost at Groix Island sometime in 1928.

Adelaide Winter, Burin, Married, 1922-24, *Beulah Mae*

Adelaide bought one share from Thomas V. Hollett, merchant at Burin, on April 15, 1922. She held this share until the vessel was sold on December 4, 1924.

Hannah Boyce, Jersey Harbour, Widow, 1926-27, *Nina Beatrice*

Daniel Boyce, merchant of Jersey Harbour, died May 5, 1926, and in his will appointed Hannah executrix. She sold her 48 shares in the *Nina Beatrice* and her partner, Orlando Bungay, sold his 16 shares to D. Boyce & Sons Ltd., undoubtedly Hannah's sons, on May 15, 1927. See Hannah also in the "Sole Owner" list above.

Abigail Horwood, Harbour Le Cou, Merchant, 1926-36-, *Syabilla M.*

Abigail's partner in the 17-ton *Syabilla M.* was John Herridge, her brother. The ship was built by Josiah Farrell of Bay de Nord, La Poile, in 1926, and it is probable that Abigail had the ship built for her purposes. In the record it states that Abigail was a fisherman and John was a merchant. This was an error, and the occupations should be reversed. Abigail's husband was Todd Horwood, merchant, and they were in the business together. She carried on the business after her husband's death. The people in Harbour Le Cou stated during my 1994 visit: "She was never a fisherman." It was said that she bought the ship for her brother Ernest Herridge, but that perhaps was because he was the master of the ship. Abigail's nickname was "Sis." John Herridge died in 1935 and Abigail became the "person acquiring title or power." As sole owner of the small ship, she had an engine installed in 1936 and continued to use the vessel for several more years. Abigail may have sold the vessel in later years since the record states: "Registry closed 16 January 1953. Vessel decayed and sank sometime ago at North Sydney, Nova Scotia. Advice received from Messrs. H. B. Nickerson & Sons. Ltd., North Sydney." It is likely that Abigail owned this vessel for 20-plus years.

Alice Maria Nash, McCallum, Married, 1933-34, *Alice M. Nash*

This 35-ton schooner had been built in 1931 by Morgan Roberts at the Head of Bay D'Espoir, and bought by Wilson Riggs, clerk, and Alice Maria Nash, both of McCallum, Hermitage Bay. Alice must have changed the name at the time of registration. They sold on November 27, 1934, to Hy. Clements Ltd., merchants at Burgeo.

Margaret Grace Ryan, St. John's, Widow, 1934-37, *Our Lady*
1934-43, *H. F. Wilson*
1934-43, *Hunter*
1934-44, *Saffron*

Daniel Ryan, merchant of King's Cove, died July 6, 1934. Executors appointed were Edmund Joseph Ryan, John Thomas McCarthy and Margaret Grace Ryan. After three years they sold all four ships to Ryan Brothers Ltd. of Trinity and three of them subsequently to other buyers. The Hon. Daniel A. Ryan had owned both firms at King's Cove and Trinity. Margaret Grace McCarthy of Carbonear had been Daniel's wife, and they had a daughter Margaret. The executor in this case was probably the wife, though she would have been in her 80s at the time.

Hannah Elizabeth Carter, St. John's, Widow, 1935-37, *Earl of Devon*

This 194-ton vessel was built at Govan, Glasgow. Alexander Carter, master mariner of Greenspond, owned 32 shares in the ship as of December 31, 1915. Alexander died on June 27, 1935, and his will of February 7, 1933, appointed Hannah Elizabeth as executor of his estate. She was by then living at St. John's. The executors for the other owner, Arthur C. Peters, and Hannah Elizabeth sold their shares on March 16 and 18, 1937, to the Commissioner for Public Utilities, Commission of Government of Newfoundland.

Isabel Frampton, White Rock, T.B., Widow, 1937-38, *Dorothy Beryl*

This 62-ton vessel was built at White Rock by Ezekiel Frampton, mariner, in 1936. Ezekiel owned 36 shares and Henry Thomas Stone, accountant of White Rock, owned 28 shares. Ezekiel died intestate on May 18, 1937, and Letters of Administration were granted to Henry Thomas who settled the estate by selling Isabel's 32 shares and his 28 shares on November 25, 1938, to Kenneth Fowlow, fisherman of Trinity East.

Miriam May Hiscock, Flat Islands, B.B., Spinster, 1938-39, *Grace Hilda*

On November 4, 1938, Miriam May bought 16 shares in the 98-ton *Grace Hilda,* which had been built at La Have, Nova Scotia, in 1916. William Hiscock, fisherman, had originally owned the vessel but before he died on July 18, 1938, he sold the vessel to William Francis Hiscock. It was from William Francis that Miriam May bought the shares. They sold the vessel on June 26, 1939.

Millicent Bragg, Glovertown, Spinster, 1940-41, *Royal Visit*

Martin Bragg, fisherman and builder of this ship, Gideon Paul and Millicent Bragg were joint owners of the *Royal Visit* as of March 15, 1940. The ship had been named to commemorate the royal visit of their majesties to Newfoundland in 1939. Millicent's sister Lily was married to Gideon Paul, and her sister Mary was the postmistress in Glovertown for many years. Millicent married late in life. The ship was lost at Horse Islands on July 15, 1941.

Mary Oldford, Salvage, B.B., Widow, 1942-43, *Britannia*

Edward Oldford, fisherman of Salvage, owned 48 shares in this ship as of June 26, 1933. The owner of the other 16 shares was Kenneth Brown, merchant of Salvage. He had a mortgage with Bowring Bros., merchants of St. John's. Edward Oldford died intestate on July 11, 1942, and Letters were granted to Reginald Brown, fisherman, and Mary Oldford, both of Salvage. They sold Mary's 48 shares on November 4, 1943, to Clarence Bertram Oldford, fisherman of Salvage.

Rosa May Pomeroy, Brigus, Widow, 1943-44, *Ellen L. Maxner*

This 105-ton ship had been built at Lunenberg, Nova Scotia, in 1899. It had been rebuilt in 1929 by Isaac Bowering of Bay Roberts for William Pomeroy and Frederick G. Pomeroy, planters of Brigus.

William Pomeroy died intestate on December 21, 1943, and Letters of Administration were granted to Rosa May and she sold her shares in the vessel on March 30, 1944, to the Monroe Export Co.

Lydia Carter, St. John's, Married, 1945-49, *R. A. Squires*

Lydia and Wilfred Eugene Butler, grocer of St. John's, were joint owners of this ship. They had the vessel altered and a new engine installed at the time of purchase. They sold the ship on November 29, 1949.

Amelia Valencia Buffett, Rose Blanche, Married, 1945-50, *Dulcie Ann*

Amelia and James Buffett, fisherman, were joint owners of the *Dulcie Ann* when they bought it in 1945. James died on June 21, 1950. Amelia moved to North Sydney and sold the vessel there on September 5, 1950. She informed the registrar at St. John's in writing. Amelia was from Diamond Cove.

Alice Hatcher, Rose Blanche, Married, 1946-49, *A. M. Hatcher*

Josiah Farrell of Bay de Nord, La Poile, built this 12-ton ship in 1932. The first owner, Robert Hatcher, fisherman of Rose Blanche, sold it on December 9, 1946, to Stanley Robert Hatcher, fisherman, and Alice Hatcher of Rose Blanche. They used the ship for three years and sold it in 1949 to Robert Allen Newman, merchant at Petites. Alice was a Cox from nearby Diamond Cove.

Harriett Chislett, Rose Blanche, Married, 1946-58, *H. B. Chislett*

Harriett Chislett and Walter Thomas Chislett, fisherman, bought this ship in 1946 from Robert Allan Newman, merchant at Petites. Harriet's husband was John, so Walter Thomas may have been her son or brother-in-law. In 1958, after 12 years of ownership, Walter Thomas bought Harriet's shares. They did not have a mortgage when it was co-owned, but when Walter Thomas bought it in 1958, he took a mortgage. Harriett was a Stroud from Rose Blanche.

Hannah Celia Hollett, Burin, Widow, 1947-48, *Golden Glow*
1947-48, *Keith V. Colin*

Thomas Vigus Hollett, Hannah's husband and merchant at Burin, died July 27, 1947. His will of August 9, 1942, did not appoint an executor. Letters of Administration were granted to Hannah Celia on March 27, 1948. Of the 26 shares that Thomas V. Hollett held in the *Keith V. Colin,* Hannah Celia sold 10 shares to Donald Hollett, 10 shares to Edward Glyndon Hollett, and 6 shares to Keith McKenzie Hollett on March 24, 1948; these were her sons. Of the 36 shares which Thomas held in the *Golden Glow*, Hannah sold 10 shares to Donald Hollett, 10 shares to Edward Glyndon Hollett, 8 shares to Keith McKenzie Hollett, and 8 shares to Hollett Sons & Co. Hannah Celia had been born at Great Burin in 1881. Her father used to hold school in their kitchen at Great Burin. Hannah and Thomas were married on January 18, 1905. They had eleven children. She died in 1949. Hannah regularly attended business meetings held at their home.

Ruth LeMoine, Rose Blanche, Married, 1947-, *Augustus Gordon*
1949-54, *Kevin J. Mildred*

Harold Llewellyn LeMoine, mariner, and Ruth LeMoine jointly owned the *Augustus Gordon* from 1947 to an undetermined date. By 1949, when they jointly bought the second ship, Harold was designated as a master mariner. Ruth was an Anderson from West Point.

Elsie May Blackwood, St. John's, Married, 1947-50, *Mary H. Hirtle*
1951-54, *Belcher Queen*

Elsie May and Job Blackwood, mariner, had left Safe Harbour, Bonavista Bay, by the mid-1940s and were living in St. John's. They jointly bought the *Mary H. Hirtle* in 1947 and took a mortgage for $8,000 with John Gilbert Higgins, solicitor of St. John's; the mortgage was discharged after four months. Then they took a second small mortgage in 1948 with Crosbie & Co. Ltd. The vessel was wrecked in Codroy Harbour on December 19, 1950. They bought the ship *Belcher Queen* in 1951 and sold it to their son Eric in 1954. Capt. Eric Blackwood was at that point an ex-RCAF pilot who in 1945 formed Newfoundland Aero Sales & Services, with seven planes delivering weekly mails to the Northern Peninsula and doing aerial seal spotting. Because of a serious plane accident on Random Island in 1952, Blackwood gave up commercial flying and thus 10 years of aerial service. See Elsie May Blackwood in the "Sole Owner" list.

Maud Billard, Harbour Le Cou, Married, 1947-53, *Dolores Kaye*
1963-84, *Austin & Dianne*

Simeon Matthew Billard, fisherman, and Maud Billard, married woman, bought the *Dolores Kaye* from Augustus and Phoebe Buckland in 1947 and sold it to U.S. subjects in 1953. It is difficult to be precise about whether the Billards owned the vessel at the time of the sale, because advice had been received from Maurice James Banfield of North Sydney, "the last operator of the vessel." That is unusual language if Banfield was the owner; it is probable that he sold the vessel on behalf of the Billards. In 1960 Simeon bought the *Austin & Dianne*, a long-liner, and in 1963 he registered it jointly in his own name and the name of his wife Maud. The ship sank on June 30, 1983, approximately 20 miles SE by E from Rose Blanche Point. It is assumed that Simeon and Maud were still the owners, because there was no transaction indicating a further sale. The period of Maud's ownership thus covered a total of 36 years. Maud was a Clarke from Harbour Le Cou; she was born on December 26, 1920, Simeon on June 4, 1918. Simeon died in 1990 and Maud was still living when I visited Harbour Le Cou in 1994.

Lilly Robins, Lumsden, Widow, 1948-50, *Gerald and Aiden*

Henry James Robins, merchant of Lumsden, died September 18, 1948, and his will, dated November 27, 1947, appointed Lilly Robins and Henry James Robins, Jr., as executors. Though probate was granted on March 8, 1949, they did not sell the vessel until February 27, 1950. The new owner was Lewisporte Wholesalers Ltd.

Ethel Maud Strickland, Rose Blanche, Married, 1948-53, *Miss Rose Blanche*

Ethel Maud and William Strickland, fisherman of Rose Blanche, bought this ship on June 7, 1948, from James Dolomount, merchant at Rose Blanche. They sold it on January 20, 1953. Ethel Maud was a Dolomount from Rose Blanche.

Annie Marie Cheeseman, Rushoon, Married, 1950-53, *Marie Cheeseman*

Joseph James Cheeseman, fisherman of Rushoon, bought this ship in 1949 and on May 16, 1950, he sold Annie Marie Cheeseman joint ownership with himself in the vessel. They sold the ship on October 27, 1953, to Charles Joseph Cheeseman and Joseph

Richard Cheeseman, fishermen of Rushoon, possibly their sons, so the vessel likely remained in the family.

Theresa Strickland, Rose Blanche, Married, 1951-53, *Justrite*
Thomas Rose Strickland, merchant of Rose Blanche, and Ralph Strickland, labourer, owned this 18-ton ship prior to 1951, and on January 30, 1951, Ralph Strickland sold his shares and rights in the ship to Theresa Strickland, Thomas' wife. They sold the ship on May 13, 1953, to a fisherman in Prince Edward Island. Theresa was a Clarke from Harbour Le Cou.

Gladys Tuck Janes, Hant's Harbour, Widow, 1952-56, *Richard J.*
Pleman Janes, merchant at Hant's Harbour, died on November 3, 1952. The two executors appointed were Lemuel Willie Janes, merchant of St. John's, and Gladys Tuck Janes of Hant's Harbour. On February 11, 1956, they sold the ship to Edwin Harrison Janes, Maxwell Carson Janes and Gladys Tuck Janes. On that same day they sold the ship to P. Janes & Sons at Hant's Harbour.

Ethel Constance Earle, Fogo, Married, 1952-, *Thomas J. Hodder*
Harold James Earle, merchant at Fogo, and Ethel Constance Earle each held 32 shares in this ship when they bought it on April 16, 1952, from Alberto Wareham Ltd. They sold it on December 7, 1953, to Earle Sons & Co. Ltd. at Fogo, and it is not known to what extent Ethel was involved in this business. Earle Sons & Co. Ltd. sold the ship in 1961.

Mary Elizabeth Hannam, Rose Blanche, Married, 1953-61, *Wilehema*
Mary Elizabeth was a King from Burnt Islands; she had come to Rose Blanche as a housekeeper and then married Charles Hannam, a fisherman. She always ensured that the ship was kept super clean, and she did work on the fish flakes. They bought this ship on February 12, 1953, from Caroline Hannam who had owned the ship since 1936. See Caroline in the "Sole Owner" list above. Charles Hannam was appointed the managing owner of the ship "under the hand of" himself and Mary Elizabeth. They sold the ship on May 30, 1961, to Marine Basin Ltd., Niagara-on-the-Lake, Ontario. After they sold the ship, Mary Elizabeth worked in the fish plant. Mary Elizabeth was still living at age 87 when I visited her in 1994.

May Hollett, Burin, Widow, 1961-62, *Golden Glow*
1961-62, *Saval*
1961-62, *Roy M*

May's husband, Donald Hollett, merchant of Burin, died on June 10, 1961. His will of October 8, 1948, had appointed May as executor. When probate was granted in April 1962 she was living at Gander. On October 16, 1962, she sold her shares in the three ships. The firm of Hollett's Sons & Co. was adjudged bankrupt as of June 13, 1961. The shares in these ships, which belonged to Hollett's Sons & Co., were sold by an executor to H. E. Senior & Sons Ltd. May Stevenson had been born at Bonavista in 1904, became a teacher, and married Donald Hollett in 1932. They had five children. May died in 1993.

APPENDIX B

Women in the Conception Bay Plantation Book *of 1805*

The *Conception Bay Plantation Book* of 1805 is of such significance for providing a context for fishing families of the east coast that an abbreviated version of it is presented here. The 1,062 properties listed here were practically all waterfront properties having a connection with the fishery, and there are several references to "by seaside," "on waterside," "beach," "room," and "dock" that confirm this. Only those items connected with a fishery are mentioned here, so all references to gardens, meadows, and so forth are omitted. However, "houses" have been included because these may have been "stores" connected to the fishery or additional dwellings for "servants" employed by a fishing family.

Most of the information presented for the 1,062 entries in the *Plantation Book* may be characterized as bare and skimpy, even skeletel. Eric Gosse in his carefully researched book on Spaniard's Bay puts flesh on the bones of the entries for Bread and Cheese Cove (an early name for Spaniard's Bay) and makes these entries throb with life and energy. To confirm this, check the six entries beginning with No. 726. Material quoted from Sean Cadigan's thesis does the same for Elizabeth Webber in No. 995. Peter Pope's thesis needs to be read to add further material of great interest.

It is true that the early court records and other similar sources present a sombre and gloomy record of some women's situations. These were rough days, and a woman sometimes had to be as rough as her crews and as bold as her contentious neighbours. But the windows into the Gosse and Smith and Barrett families that Eric Gosse provides, give us a very different picture – of women who were hard-working, civilized, involved in community affairs, caring for extended families, yet bold and even fierce in their determination to ensure the family's survival. It is hoped that other writers will do for their communities what Eric Gosse has done for Spaniard's Bay.

To provide a better idea of what a "plantation" or "room" was like I quote from the *Evening Telegram* of October 4, 1919:

> For Sale in a Northern Outport.
> 4 cod traps with Cod Bags, Anchors and Grapnels.
> 1 motorboat – will carry 20 qtls. fish.
> 1 large fishing stage – can salt 2,000 qtls. fish
> Flake Room for 100 qtls.
> 1 large shop with store – can store 1,000 qtls. dry fish
> Fish Beams, Scales, etc.
> Part Oil Factory, with puncheons.
> Baccalieu Tickle only 3 miles away.

That would be a reasonable description of what these Conception Bay plantations were like about the year 1800.

These are the entries in the *Plantation Book* that deal with women:

No. 2: Sarah Banks, Bay de Verds, two flakes, reference to "beach." Sarah had received this property "by her father-in-law's will." Date of possession 1716. It was being leased for 20 years at 4 pounds sterling per year.

No. 12: Joanna Vinsen, Bay de Verds, formerly a "Ship's Room," i.e., where large ships from Europe tied up. Joanna sold the property to William Lockyer for 2 pounds. Date is 1774.

No. 63: Ann Pendergrast, Lower Island Cove. Ann had purchased this property from Thomas Kenelly for 5 pounds in 1796.

No. 64: John and Sarah Wells, Lower Island Cove. John and Sarah sold this property to William Broderick for 3 pounds in 1804.

No. 65: Mary Lowe, Lower Island Cove. John Lilly had purchased this property from Mary Lowe. Price not given. William Barrett was the occupier. Date is not given.

No. 69: Frances Hiscock, Lower Island Cove. This property is noted to be "by seaside." It was given to George Gaudin, Jr., "By deed of gift from Frances Hiscock." It had one stage and one flake. Date is not given.

No. 70: Mary Dimond, Lower Island Cove. Mary received this property "By deed of gift from her father," George. It had one flake. Date is 1803. Her father had a separate piece of property in Lower Island Cove.

No. 74: Ann Snelgrove, Lower Island Cove. Ann received this property "By deed of gift from her father-in-law." It had one-third of a stage, one flake and three houses (probably one or two of these were "stores," used in connection with the fishery, or they may have been for house servants employed in the fishery).

No. 75: Elizabeth Lock, Lower Island Cove. This property had been "Bequeathed to Elizabeth Lock by her grandfather's will." In 1770 it was in possession of John Lock. It had one-third of a stage, one flake and four houses.

No. 76: Elizabeth Lock, Lower Island Cove. In 1757 it was in possession of Charles Normore by deed of gift from Elizabeth Lock. It had one stage, one flake and two houses.

No. 79: Ann Snelgrove, Lower Island Cove. This was Ann's second piece of property, purchased from John Phillips for 9 pounds. It had one-half stage and one flake. A comment in the record is of interest: "…part of meadow formerly in dispute between Thomas Cooper and her admeasuring 72 yds. now divided equally by consent of both parties and staked out by me this 6 of May 1806 to satisfaction of both. Charles Carroll." Date is 1756.

No. 95: Ann Snelgrove, Lower Island Cove. This is the third piece of property connected with Ann Snelgrove. She and her two partners

(E. W. and T. R.) sold this property to John Lilly for 2 pounds. Date is 1806.

No. 96: Catherine Walsh, Lower Island Cove. Catherine sold this property to Robert Johnson for 5 pounds. Date is 1798.

No. 98: Ann Snelgrove, Lower Island Cove. Ann's fourth property! Purchased from George Moores. Date is 1802.

No. 105: Widow of William English, Devil's Cove. Property owned by John, James, Richard and Thomas English with the widow of William English. It had one stage and one flake "and all the Droke." Date is 1796.

No. 122: Mary Dunphy, Northern Bay. Mary had sold this property to Edward Butler for 13 shillings 6 pence. Date is 1790.

No. 135: Sarah Bolan, Oaker Pits Cove. Sarah had received this "By deed of gift from her father." She was leasing it to Jon Parsons for 20 shillings per year. Date is not given.

No. 136: Wife of Charles Webister, Oaker Pits Cove. This property was listed in Charles Webister's name but it had been "bequeathed to his wife by R. Wells' will." Date is 1801.

No. 169: Mother of Michael Connell, Western Bay. This property was in possession of Michael Connell "by deed of gift from his mother." Date is 1786.

No. 174: Diana Eels, Western Bay. This consisted of two plots. Date is 1788.

No. 181: Mary Dalton, Western Bay. This had one stage, one flake and three houses. It was purchased from James Thomas for 150 pounds. Date 1784. "Note: Mary Dalton is to enjoy this estate during her natural life then 'tis to devolve to nine brothers 2 of which named John Dalton and Matthew Dalton are excluded from their parts in consequence of their Father's will." See also No. 254.

No. 182: Mother-in-law of William Whiteway, Western Bay. This came into possession of William Whiteway "by his mother-in-law's will," but was noted to be "void" in 1805. It was claimed by Noah Perry.

No. 187: Grace Delaney, Western Bay. Noted to be "bounded on waterside." Date is 1748.

No. 202: Mother of William Crocker, Bradley's Cove. This property was bequeathed to William Crocker "by his mother's will." It had one-half stage, one flake and four houses. Date is 1796.

No. 204: Mother of William Dimond, Adams Cove. This was "bequeathed by mother's will" to William Dimond & Co. The two plots had one stage, one flake and seven houses. Date is 1796.

No. 207: Mother of "The Martins," Adams Cove. This came into possession of the Martins "by deed of gift from their mother's will." It was being rented by John Murray for one shilling per year. Date is 1795.

No. 208: Mother of H. and Stephen Luther, Adams Cove. This property was "bequeathed by their mother's will." It had one stage, one flake and five houses. Date is 1790.

No. 214: Mother-in-law of Joseph Bagg and Michl. Cahill, Adams Cove. This property came to the two men "by their mother-in-law's will." Date is 1797.

No. 215: Mother of Philip Holmes, Adams Cove. This property came to Philip Holmes "by deed of gift from his mother." Date is 1799. See John Holmes No. 221.

No. 216: Grandmother of William Gill, Adams Cove. This property came to William Gill "by his grandmother's will." It had one-third stage, one flake and three houses. Date is not given.

No. 219: Mother-in-law of Robert Evans, Adams Cove. This consisted of two plots and had one-third stage, one flake and two houses. The plot with the flake and stage came to Robert T. Evans "by his mother-in-law's will." Date is 1799.

No. 220: Mother-in-law of John Rines, Adams Cove. This property had one-third stage and one flake and came to John Rines "by his mother-in-law's will." Date is 1756.

No. 221: Mother of John Holmes, Adams Cove. This consisted of two plots and came to John "by deed of gift from his mother." Date is 1791. John may have been the brother of Philip Holmes (see No. 215), so the mother may have had sufficient means to provide for both of her sons separately.

No. 222: Mother of William Crocker, Adams Cove. This property had one-half stage, one flake and three houses and came to William Crocker "by his mother's will." It was being rented to Jas. Adams and Patrick Murphy for 3 pounds per annum. Date is 1796.

No. 229: Mother-in-law of Thos. Butt & Edward Lawrence, Blackhead. This property had one stage, one flake and three houses and came to the two men "by deed of gift from their mother-in-law." Date is 1790.

No. 231: Mother-in-law of John Moran, Blackhead. This property had one stage and one flake and came to John "by deed of gift from his mother-in-law." The present occupier was Dinah Pippy; was she involved in the fishery? Date is 1806.

No. 240: Tabitha Mulcahy, Broad Cove. This property consisted of two plots and had one stage, one flake and three houses. It came to Tabitha "by her husband's will for her natural life then to devolve to Bakers children." Tabitha was occupying the property in 1750. In 1766 it was occupied by David Whealon.

No. 245: Widow of John LeGrow, Broad Cove. This property had one flake and three houses, and came to the five children of John LeGrow "by deed of gift from their mother." It was occupied in 1776 by Joseph Bagg.

No. 250: Mother of Michael Thistle, Mully's Cove. This property had one stage, one flake and two houses. It was "bequeathed to (Michael) by his mother's will." Date is 1777.

No. 254: Mary Dalton & Sons, Broad Cove. This is very probably the same Mary Dalton as in No. 181. This was "bequeathed by her father's will" and in 1777 was being rented to Morgan Hingay for 10 shillings per annum.

No. 287: Catherine Bryan, Chappels Cove. "This room is being claimed by Catherine Bryan, daughter to Jno. Terry the original owner." It had one stage, one flake and two houses and in 1795 was being occupied by Philip Whealon. See also No. 395.

No. 295: Grandmother of Michl. Corbut, Chappels Cove. This was left to Michael "by deed of gift from his grandmother." Date is 1801.

No. 298: Bridget Kennedy, Harbour Main. This had been held by Bridget and was purchased from her by Wm. Moores, Jr., for 6 pounds 10 shillings. Date is 1803.

No. 301: Mary Wall, Harbour Main. This property which Mary owned had been "cut out of the woods." Date is 1794. James Wall owned an adjacent property with one stage, two flakes and three houses. Date is 1786. Was this a husband and wife team?

No. 324: Julianna O'Neal, Harbour Main. This was "by seaside" and had one stage, one flake and three houses, and had been purchased from Philip Eno or Shano (writing is indistinct). See No. 338. Date is 1793.

No. 329: Ann Penny, Harbour Main. Ann was occupying this property which had been left to her "by her grandfather's will." Date is 1794.

No. 334: Mary Drohan, Harbour Main. This property had been "cut out of the woods" and was occupied by Mary. Date is 1792.

No. 338: Elizabeth Terry, Harbour Main. This was listed under the names of Thomas, John, Wm. and Edward Terry and had one stage, three flakes and two houses. It had been purchased from Ph. Eno (same as Philip Eno or Shano in No. 324?) "for 1 pair of shoes," and was being occupied in 1774 by Elizabeth. Whether the four brothers or the sister made this astute deal is uncertain.

No. 351: Mother-in-law of Jas. Hedderson, Gasters. This was left to Jas. Hedderson of Turk's Gut "by deed of gift from his mother-in-law" and he was occupying it in 1796.

No. 382: Mary Merrigan, Collier's Dock. This property was unoccupied in 1780 but was owned jointly by Edward and Mary Merrigan; it had been "cut out of the woods."

No. 390: Mother-in-law of John Wall, Collier's. "Note: This room is disposed of by Bill of Sale dated 1 December 1787 to Thos. and Lawrence Dunn for 90 pounds." Apparently it was later than this that it came to John Wall "by deed of gift from his mother-in-law" for he was occupying it in 1801.

No. 392: Mary Whealon, Mary Cullin and Mary Cole, Collier's. This property, which had one stage, one flake and two houses was curiously divided among five people: Mary Whealon was to hold 44 yards "during her natural life," Jas. Cole 43 yards and John Cole, Mary Cullin and Mary Cole 15 yards each, "in the whole 132 yds." The property came to them "by deed of gift from John Cole"; this was probably John Sr. dividing his room among family members. Date is 1778.

No. 394: Ann, Catherine and Wm. Narrowcot, Collier's. This had one stage, two flakes and one house, and came to the three "by deed of gift from their grandfather." In 1781 it was occupied by John Lee. Was he the husband or brother of Mary Lee? See No. 395.

No. 395: Catherine Bryan and Mary Lee, Collier's. This property had come to the two women "by deed of gift from their father-in-law." It had one stage, one flake and two houses and was occupied by Peter Bryan in 1793. Was this Catherine's husband? For Catherine's other property see No. 287. "Note: Mrs. Hedderson is to have the grass annually that the meadow ground produces during her natural life." This may have been Mrs. Jas. Hedderson, whose husband held property in Gasters (See No. 351), Collier's and Turk's Gut. The name is also spelled "Edderson." See also No. 394 for John Lee.

No. 405: Ann Rober (or Hober), South Shore, Broad Cove. This had been purchased from Ann by John Hanall of Portugal Cove for 25 pounds. Date is 1797.

No. 454: Sisters of Wm. Mugford, Port de Grave. This property had one stage, two flakes and two houses, and was "bequeathed by father's will" to William Mugford and two sisters (unnamed). A note says it was "built" in 1770.

No. 457: Wife of George Voy, Port de Grave. These two plots were given or left to George Voy "by his wife." It had been "built" in 1802 and was then being occupied by Patt. Brown who was leasing it for 10 years at 6 pounds (per annum?).

No. 459: Wife of Samuel Batton, Port de Grave. This "was bequeathed to his wife by her father." Date is 1789.

No. 474: Mother of Thos. Snow, Port de Grave. This was "bequeathed by his mother's will." Date is 1777.

No. 475: Exactly the same as No. 474.

No. 484: Mary Butler, Port de Grave. This extensive "room" of 132 yards frontage consisted of one wharf, two flakes, four gardens, four stores, one cook room and four houses. It came to Mary "by deed of gift from Edward Baker." John Butler (Jno. Butler's son was undoubtedly Mary's brother. See No. 492, also No. 485). Jos. Furneaux, agent to Wm. Newman, is mentioned, probably because part of the "room" was being leased to Wm. Newman for 11 years at 45 pounds per annum. Date is 1765.

No. 485: Mary Butler, Back Cove, Port de Grave. Similar to No. 484, except that M. Butler (presumably Mary) is the occupier. Date is 1765.

No. 491: Mother of Jas. and Wm. Wells, Port de Grave. These two plots had one stage, one flake, one dwelling house and one store, and came to the two brothers "by deed of gift from their mother." Date is 1776.

No. 492: Mary Butler, Port de Grave. This was "bequeathed to her by her father's will, Jno. Butler." It was being leased to Wm. Newman and Wm. Jeffers. Date is 1765. See also Nos. 484 and 485.

No. 493: Ann Mugford, Port de Grave. Her first name looks like "Amy" in the record; it is indistinct. The property had one stage, two flakes and three houses and had been "cut and cleared agreeable to Act Wm. III 25.7." Ann's husband Robert left his property to her "for her natural life" in his will of 1793. Ann died in 1813. See Chapter 2 under "Women and Property" for the complicated situation that arose for the Mugfords, because of Ann's inappropriate actions. The date for No. 493 is 1725, which is probably when the property originally was granted to the Mugford family.

No. 504: Charity Noseworthy, Patience Dwyer and Edward (?) Nugent, Port de Grave. "Edward" may be "Edwina." Consider the notes. This property had been "possessed by their ancestors for the term of 105 years." Another note says: "Purchased from Ann Stretch for 20 pounds by their father who died intestate and now claimed by the 3 daughters." Ann Stretch must have been a family member, one of the "ancestors." Ann Stretch also owned two properties in Harbour Grace. The occupier of No. 504 in 1798 was Ed. Nugent (Edwina?). See also Nos. 978 and 1036.

No. 506: Ann Coveduck, Port de Grave. Ann's father, Matthew Anthony, left Ann a piece of property which had been "possessed for 105 years" by their ancestors. Date is 1780. He also left Ann's husband, John Coveduck, a slightly smaller piece of property which had been in the family for 105 years, so he had equally divided his property between his daughter and his son-in-law. Curiously the date on No. 505 (John Coveduck's property) is 1760. The property had been in the Anthony family from at least 1700.

No. 514: Mother of George Dawe, Sr., Port de Grave. This property, which had one stage, one flake and three houses, was "possessed by his ancestors for 160 years," and was "bequeathed to him by his mother's will." Date is 1755. Assuming that the person who recorded the data was figuring backward from 1805, this property had been in the Dawe family since 1645.

No. 528: Bridget Morgan, Port de Grave. This had one stage, two flakes and two houses and came to Bridget "by deed of gift from her brother-in-law." It had been "possessed for 30 years." Bridget was occupying the property in 1785. Joseph Morgan owned and occupied the property that was "bordering on Bridget Morgan" (No. 529). It had one stage, three flakes and three houses. This would seem to be another brother (or brother-in-law) and sister (or sister-in-law) team. There is no date on Joseph Morgan's property.

No. 539: Mother of Richard Porter (Mary?), Hibb's Hole. Richard Porter's property in No. 539 has a note "bounded on the S. W. by his Mother's Room..." For possible explanation, see No. 540. See also No. 605.

No. 540: Mary Porter, Hibb's Hole. This property had one stage, one flake and two houses and was owned by Richard Porter, having been "purchased from his mother for 6 pounds per annum during her life time." Mary Porter was occupying the property at the time of recording. No date given. The property had been "possessed by his ancestors for 106 years," i.e., from 1699. See also No. 605.

No. 544: Rachel Dawe, Port de Grave. Rachel's husband had been Jno. Dawe. This property had been "bequeathed by her father's will." She was occupying the property in 1775.

No. 547: Widow King & Sons, Brigus. This had one stage, one flake and one house. It had been purchased from Conway Heighington for 26 pounds and had been "possessed for 21 years." Conway Heighington held property in Harbour Main and Susan Heighington of Bristol bought property in Musketta (See No. 807). No. 547 was being occupied by Widow King in 1784.

No. 571: Dinah Roberts, Brigus. These two plots were "by Catherine Antle's garden" and had one stage, two flakes and one house. It had been "possessed by their ancestors for 80 years" and came to Dinah "by deed of gift from her father." Dinah was occupying the property in 1771. See No. 575 for Diana Roberts. See also No. 574 for Widow Antle.

No. 572: Mother of John Rabbitts, Sr., Brigus. This property had been "possessed for 55 years" and came to John Sr. "by deed of gift from his mother." Apparently his nephew, John Rabbitts, Jr., was his partner.

No. 574: Widow Antle and Wm. Antle, Brigus. Both Wm. Antle and Widow Antle occupied the property in 1782 which had come to the widow "by deed of gift from her father-in-law." Wm. Antle, Sr., also held property in Brigus (No. 567) which had been "bequeathed to him by his father-in-law's will."

No. 575: Diana Roberts, Brigus. Whether this is the same as Dinah Roberts (No. 571) cannot be determined; both names are clearly spelled. This property was owned by John Roberts (Diana's son) "by deed of gift from his mother." Date is not given.

No. 579: Grace Walker, Brigus. This property had come to Grace "in consequence of lying void." She was occupying it in 1797.

No. 582: Grace Norman, Brigus. This property came to Grace "by deed of gift from her mother-in-law" and she was occupying it in 1786.

No. 583: Ann Roberts, Brigus. Part of this property was purchased, part came to Ann "by deed of gift from her father," and part was "cut out of the woods." Ann was occupying the property in 1772.

No. 587: Mother of Robt. Knight, Brigus. This property came to Robert by "deed of gift from mother," and he was occupying it in 1800.

No. 600: Mary Anthony, Cupids. Mary acquired this property because it was "lying void." She was occupying it in 1770. Joseph Anthony owned No. 604 "by deed of gift from his father-in-law." Was he Mary Anthony's husband?

No. 605: Mary Porter, Cupids. Wm. Coveyduck purchased this property from Mary Porter for 4 pounds, and he was occupying it in 1783. See Nos. 539 and 540.

No. 608: Mary LeDros, Cupids. This property was "cut and cleared agreeable to Act. Wm. III chap. 25. Sec. 7." Mary was occupying it in 1762.

No. 676: Widow Merser, Bay Roberts. This property had been "possessed for 100 years" and belonged to "the heirs (children) of Jas. Merser Sr. (deceased)." It was occupied by Widow Merser in 1800.

No. 682: Patience Parsons, Bay Roberts. Patience had become owner of this property as a "gift from her father." She was occupying it in 1769. See also No. 963 for her property in Harbour Grace.

No. 689: Grandmother of Edward Russel, Bay Roberts. This property had come to Edward "by his Grandmother's will." Date is not given.

No. 696: Mother-in-law of Thos. Caravan, Bay Roberts. This property had come to Thomas "by his mother-in-law's will," and he was occupying it in 1781.

No. 697: Mother of Wm. Snow, Bay Roberts. This property came to William as a "gift from his mother." He was occupying it in 1795.

No. 699: Mother of Wm. & Edward French, Bay Roberts. This property was "a gift from their mother" to William and Edward. They were occupying it in 1780.

No. 726: Sarah Barrett, Bread and Cheese Cove (Spaniard's Bay). This large plantation measured 187 yards by 200 yards and had one stage, two flakes, two houses, three gardens and one meadow. It was "cut and cleared" and granted to Sarah Barrett in 1755. Eric Gosse affirms that Sarah Barrett owned "this important fishing establishment," this "large and thriving fishing enterprise," and that she was probably related to Abram and David Barrett and "may have held the power and influence as matriarch in this family." Of the 38 plantations in Spaniard's Bay, Sarah's was designated as No. 24.

No. 728: Sisters of Thos. Smith, Bread and Cheese Cove. This property came to Thomas Smith "by deed of gift from his sisters." Thomas was occupying it in 1797. Eric Gosse affirms that it consisted of one stage, two flakes, two houses and gardens, and was "one of the best establishments in Bishop's Cove." It was designated as No. 26 of the Spaniard's Bay plantations.

No. 730: Jane Smith, Bread and Cheese Cove. This plantation consisted of more than eight acres and had been registered by Jane in 1763. Part of it had been "cut and cleared" and part had been purchased from Jno. Percey for 10 pounds. There was one stage, one flake, one house and three gardens. Eric Gosse writes: "It was considered to be a typical well-balanced fishing establishment for a family engaged in the inshore fishery. The cash returns from the fishery, combined with fresh vegetables from its gardens and meat from raising sheep and goats, ensured a simple and healthy livelihood for Jane Smith and her dependents. How this remarkable woman came to own such a well-developed fishing enterprise is unknown. She had to manage and control her rough Dorset fishing crew, supervise the household chores and the cultivation of her fields. Also she was responsible for the proper salting, drying and marketing of the all-important cod as well as the pickling of other species, such as caplin, herring and mackerel."

No. 734: Children of Richard Gosse, Bread and Cheese Cove. This property belonged to "the children of Richard Gosse deceased," but the language of the next sentence is curious: "Bequeathed to her husband by his father's will." Date is 1775. According to Eric Gosse, "These orphans were brought up by their great-uncle Robert Gosse, Sr., and his wife Catherine."

Plantation No. 33 in Spaniard's Bay, Catherine Gosse, Bread and Cheese Cove. Robert Gosse, Sr., died in 1794, and his widow, "the redoubtable Catherine," took over the fishing establishment, though there is no record of this in the *Plantation Book*. Eric Gosse writes: "She proved herself to be a successful businesswoman and in 1804 she purchased a sizeable plantation in Carbonear from Thomas Mann for 30 pounds sterling. This property was immediately leased to the local firm of Gosse & Chancey for 31 years at the rate of 5 pounds sterling per annum. As a widow and dowager, Catherine was the undisputed matriarch of the Gosse family of Spaniard's Bay until her death in 1818." She had reached the age of 92. See also No. 892.

Not in *Plantation Book*: Sarah (Sheppard) Gosse, Bread and Cheese Cove. Sarah's husband Israel operated fishing plantations with his brother Moses in Bishop's Cove, Northern Cove and Harbour Grace. Israel died in 1876, and Sarah outlived him by eight years. Eric Gosse writes: "...after which the original fishing plantations were sold and all the children of the marriage moved to Northern Cove." It is not known to what extent Sarah was involved in the business; it is probable that she continued as overseer in the mode of other women of Spaniard's Bay.

No. 742: Mother-in-law of Michael Power, Spoon Cove. This undescribed property was "a gift" to Michael "from his mother-in-law." Date is not given.

No. 745: Elizabeth Smith, Island Cove. Elizabeth Smith had owned this property but sold it to Joseph Drover for 5 pounds. Joseph was occupying it in 1776.

No. 757: Elizabeth Lynch, Island Cove. Elizabeth obtained this property by "gift from her father-in-law." She was occupying it in 1789.

No. 768: Ann Noseworthy, Bryant's Cove. Ann received this property by "Grant from the Governor." She was occupying it in 1801.

No. 776: Elizabeth Higgins, Harbour Grace, South Side. This property had been "cut and cleared" and was occupied by Elizabeth in 1774.

No. 788: Susannah Taylor, Musketta, South Side. Susannah received this property because it had been "lying void." She was occupying it in 1779.

No. 807: Susannah Heighington, Musketta, North Side. Susannah's residence was given as Bristol, so she may have been a recent arrival, or an absentee landlady. She purchased the plantation from Samuel Pike for 150 pounds in 1803 and it was being occupied by D. Hartry and M. Kerney of Harbour Grace. These may have been her agents. See Nos. 305 and 547.

No. 812: Elizabeth Pippy, Musketta, North Side. Elizabeth Pippy of St. John's owned this property and it was being occupied by P. Carew of Musketta in 1796. It was being leased for several years at the rate of 1 pound 10 shillings per annum.

No. 814: Patience Pynn, Musketta, North Side. Patience Pynn of St. John's obtained this plantation in 1765, part of it "by deed of gift from Wm. Pynn" and part purchased from Jno. Butt for 15 pounds.

No. 819: Ann Wells or G. & S. Kemp & Co., Carbonear, South Side. This must have been a joint ownership although "or" is used, not "and." It had been "cut and cleared" and was being occupied by Ann in 1786.

No. 820: Mary Buckingham, Carbonear, South Side. Mary had received this plantation "by her former husband's jointure." "Jointure" is a term used in law and refers to an estate settled before marriage. Was this similar to the secular marriage contract in Quebec that was additional to religious marriage and separated the wife's assets from those of her husband? The plantation was being occupied by Jno. Buckingham in 1775.

No. 822: Elizabeth Pike, Carbonear, South Side. Elizabeth was renting this property to Jos. Crimp in 1799 for 8 pounds per annum. See also Nos. 856, 862, and 904.

No. 826: Mary Keefe, Carbonear, South Side. This property had been "cut out of the woods" and was being occupied by Mary in 1765.

No. 830: Frances Thistle, Carbonear. Frances had inherited this property "by her husband deceased." She was occupying it in 1772.

No. 849: Susannah Enger, Carbonear. Susannah's surname is indistinct in the record. She had sold this property to Thomas Reynolds for 5 pounds. Date is not given.

No. 856: Elizabeth Pike, Carbonear. This property was purchased by Elizabeth from Richard Diffix for 120 pounds, and she was occupying it in 1715. This must have been another Elizabeth Pike. Note the difference in dates from No. 822. See also Nos. 862 and 904.

No. 862: Elizabeth Pike, Carbonear. This property was purchased by Elizabeth from Edward Sweetland for 9 pounds. Date is not given. See also Nos. 822, 856 and 904.

No. 873: Mary Williams, Carbonear. Michael Walsh purchased this from Mary Williams for 10 pounds. Date is not given.

No. 882: Mary Ferrers, Carbonear. Mary was leasing this property from William Parsons for 2 pounds per annum. See Nos. 885 and 886. Date is not given.

No. 885: Frances Tucker, Carbonear. Frances had two plots. One she was leasing to Chs. McCarthy for 8 pounds per annum; the other she was leasing to Mary Ferrers for 5 pounds per annum. See Nos. 882 and 886. Date is not given.

No. 886: Mary Ferrers, Carbonear. This plot was "cut out of the woods" and Mary was occupying it in 1785. See Nos. 882 and 885. Mary owned one property and was leasing two others. She was a busy woman.

No. 887: Mother of Wm. Howell, Carbonear. This property came to William Howell as a "gift from his mother." Date is not given.

No. 888: Grandmother of Jno. Robt. & Wm. Howell, Carbonear. This property, which both brothers were occupying in 1800, had been "a gift from Grandmother."

No. 889: Mother of Jas. Howell, Carbonear. This property, which was being occupied by James Howell in 1775, was a "gift from his mother."

No. 892: Catherine Gosse, Carbonear. "The redoubtable Catherine" of Bread and Cheese Cove purchased this plantation from Thomas Mann for 30 pounds and was leasing it to Gosse & Chancey for 31 years at 5 pounds per annum. See the special entry after No. 734 for Catherine Gosse.

No. 893: Ann Toque, Carbonear. This property was "bequeathed by their father's will" from the late Robert Howell to his four children, William, Jno. and Peter Howell and Ann Toque. It was owned by all four of them in 1800. In 1831 and 1835, as the widow of Philip Toque, Ann was a partner with Simon Levi in the two ships *Alpha* and *Elizabeth*. See her in Chapter 3 of this study.

No. 904: Elizabeth Pike, Carbonear. This is the fourth piece of property registered under the name of Elizabeth Pike. This one had been purchased from Samuel Pike and in 1780 was occupied by Charlotte Pike and Samuel Lilly. See Nos. 822, 856 and 862.

No. 905: Elizabeth Hayward, Carbonear. This property had been "cut out of the woods" and was occupied by Elizabeth in 1769.

No. 906: Mother of Henry Knight, Carbonear. This property came to Henry Knight as a "gift from his mother." He was occupying it in 1786.

No. 907: Mary Evory, Carbonear. Mary purchased this property from William Penny for one whipsaw worth 45 shillings. An astute deal! She was occupying the property in 1787.

No. 914: Elizabeth Knight, Carbonear. This property came to Elizabeth by "deed of gift from her father." Jno. Warn and B. Linthorne were occupying it in 1755, so apparently she was leasing it.

No. 917: Elizabeth Knight, Carbonear. Elizabeth had owned this property, but by 1797 Callaghan McCarthy had purchased it from her for 8 pounds.

No. 919: Mother of Robert Penny, Carbonear. This property came to Robert as a "gift from his mother." Date is not given.

No. 938: Sarah Squires, Freshwater. Sarah had purchased this property from Richard Dolby and Jno. Sweet for 55 pounds, and was leasing it to Ed. Cahill for 11 years for 2 pounds 15 shillings (per annum?). Date is not given.

No. 962: Frances Marshall, Harbour Grace, North Side. Jonathan Parsons purchased this property from Frances Marshall for 10 pounds in 1770. See also No. 974.

No. 963: Patience Parsons, Harbour Grace, North Side. This property had "descended to her late husband by his father" and was owned by Patience in 1765. See also No. 682 for her property in Bay Roberts.

No. 964: Mary & Wm. Balding, Harbour Grace, North Side. This property had been "cut out of the woods" and was occupied by Mary and William in 1765.

No. 974: Frances Marshall, Harbour Grace, North Side.This property was a "gift from her father" and Frances was occupying it in 1775. See also No. 962.

No. 977: Widow Kennedy, Harbour Grace,. North Side. This property was a "gift from her father" and the widow was occupying it in 1770.

No. 978: Ann Stretch, Harbour Grace, North Side. Abraham Parsons had purchased this property from Ann Stretch for 10 pounds. Date is not given. See Nos. 504 and 1036. Ann had also owned property in Port de Grave.

No. 982: Grandmother of Hy. Pynn, Harbour Grace, North Side. This property had been left to Henry Pynn of England "by his grandmother's will." It was being leased, so it is not certain whether Henry ever occupied the property. Date is not given.

No. 987: Jane Cooke, Harbour Grace, North Side. This property had come to James Cowan "by Jane Cooke's will." This is very probably the same "Jane Cook" to whom Sean Cadigan refers in his thesis: "Jane Cook, who acted for (Ann) Protch in the matter (of leasing properties) also handled their account with merchant James MacBraire and rented other property at Harbour Grace to John Clements." Jane Cooke was typical of the kind of women who, as Cadigan asserts, "managed property and accounts left to them by their husbands." Date for No. 987 is not given.

No. 988: Amy Thistle, Harbour Grace, North Side. Amy inherited this property "by her Father-in-Law's will." She was leasing it to four tenants for an undisclosed number of pounds per annum. Date is not given. See also Nos. 994, 1037 and 1047. Amy was the owner of four properties.

No. 989: Mary & Ann Moore, Harbour Grace, North Side. Mary and Ann had inherited this property as a "deed of gift from their grandfather." It was being leased to Dr. John Sterling. Date is not given.

No. 991: Mother of James Sheppard, Harbour Grace, North Side. Curiously, the only note in this entry is "deed of gift from her mother" which means that the property came to James through his wife, or this is an error. No date or other information is given.

No. 994: Amy Thistle, Harbour Grace, North Side. Amy received this property "By permission from Governor Byron." Date is not given. See also Nos. 988, 1037 and 1047.

No. 995: Elizabeth Webber, Harbour Grace, North Side. The only note in this entry says "buildings burnt last year," i.e., 1804 (?). This undoubtedly was the same Elizabeth Webber referred to in Sean Cadigan's thesis: "…Mrs. Thomas Thirts of Harbour Grace tried to protect her claim to a fishing room from the intrusion of a neighbouring planter family: Elizabeth Webber & Sons. The court allowed Webber – a widow left with eight children to care for – to build her stage on the property. She had to contend with her eldest son Charles trying to mortgage their property to a merchant Thomas

Lewis to pay his own debts, but the court allowed Charles only an eighth part of the total property, and gave Mrs. Webber one third of her late husband's land as her sole property." Life was hard. Date for No. 995 is not given. See also No. 1007 for another property which Elizabeth Webber had sold.

No. 1000: Mary Martin, Harbour Grace, North Side. The only note in this entry says: "Descendants of Joseph Martin, Widow Mary Martin, by her father-in-law's will." It may be assumed that Mary Martin was occupying the property with her children. Date is not given.

No. 1001: Mary Andrews, Harbour Grace, North Side. This property came to Mary "by her uncle's will." It had been in possession of the family for 91 years, i.e., since 1714. In 1780 it was being rented to Robert Holden.

No. 1003: Charlotte Pike, Harbour Grace, North Side. Charlotte had purchased this property in 1760 from Thos. Terry & Capt. Messervey for 90 pounds. It was being leased to Wm. J. & Matt. Stevenson of Harbour Grace for fifteen years at the rate of 15 pounds per annum. See also Nos. 904 and 1043.

No. 1007: Elizabeth Webber, Harbour Grace, North Side. Elizabeth sold this property to Henry Thomas and he was occupying it in 1765. See also No. 995.

No. 1008: Grandmother of John French, Harbour Grace, North Side. This property came to John by "gift from his grandmother," and he was occupying it in 1775. See No. 1020.

No. 1011: Mother of George Andrews, Harbour Grace, North Side. This property had been in possession of the Andrews family for 30 years and came to George as a "gift from his mother." Date is 1802.

No. 1020: Widow French, Harbour Grace, North Side. This property was termed "French's Plantation." It came to Widow French "in consequence of lying void." She was occupying the property, but a date is not given. See No. 1008.

No. 1036: Ann Stretch, Harbour Grace, North Side. This property had come to Jas. Cowan "by Ann Stretch's will." See also Nos. 504 and 978.

No. 1037: Amy Thistle, Harbour Grace, North Side. This property was "cut out of the woods" and Amy was occupying it in 1785. See also Nos. 988, 994 and 1047 for Amy's other properties. She had achieved the distinction of owning four properties, at least two of which were being leased to tenants.

No. 1040: Honor Lynch, Harbour Grace, North Side. This property was "cut out of the woods" and Honor was occupying it in 1795.

No. 1043: Charlotte Pike, Harbour Grace, North Side. There is no other information on this property. See also Nos. 904 and 1003.

No. 1047: Amy Thistle, Harbour Grace, North Side. Amy received this property "bequeathed by her father-in-law's will." In 1750 she was renting it to "sundry persons," in total eight tenants.

No. 1050: Jane Singeon, Harbour Grace, North Side. Jane received this property "by her mother's will." Jane was occupying it in 1783.

There were other women owners of plantations whose names do not appear in the *Plantation Book*. One of these was Catherine Thistle who is mentioned in Chapter 3 in the section on "The Eighteenth Century." From court records it is noted that in 1757 Catherine went to court and received legal action against Jersey men who were encroaching on her plantation at Harbour Grace. In the same year, Mary Dinn of Musketta sent a petition to the governor of the day stating that her family had an established right dating back more than 50 years to harvest grass on Carbonear Island for her animals. The commander of the fort at the island had refused Mary permission, but the governor overruled the commander, and Mary was allowed to continue the practice. No doubt there were many others.

There are 1,062 entries in the *Plantation Book*. A total of 147 women are mentioned in connection with 156 of these plantations. In addition to the entries listed previously, there were several plantations that were left to men through their father-in-law's will, uncle's will, and so forth.

Thus a total of 163 women (and possibly 172 properties) may be involved, if all these wives were still living. The prevailing dominance of the patriarchal line would account for inheritance through the wife's husband. In fact, it seems surprising that there were not more of these in the records. There are a total of 16:

No. 16: Bay de Verds. Daniel McCarthy, Thos. Lynch and Jas. Blunden inherited this property "By deed of gift from their wives' uncles." Date is not given.

No. 147: Oaker Pits Cove. Richard Halfyard received a plantation with one stage, two flakes and four houses in 1761 "By deed of gift from his father-in-law."

No. 555: Brigus. This plantation came to John Morgan, "bequeathed by his father-in-law's will" in 1805.

No. 556: Brigus. John Sparks received a plantation with one stage, four flakes, one wharf, and four houses in 1785, "bequeathed to him by his father-in-law's will." It had belonged to the family for 60 years.

No. 567: Brigus. Wm. Antle, Sr., had a plantation "bequeathed to him by his father-in-law's will." Date is not given, though from No. 574 we may deduce that it was about the 1780s.

No. 588: Brigus. John Plowman received this plantation as a "deed of gift from his father-in-law." Date is not given.

No. 604: Cupids. Joseph Anthony inherited this plantation "by deed of gift from his father-in-law." Date is not given. See No. 600 for a possible connection to Mary Anthony, in which case it would be about 1770.

No. 659: Bay Roberts. Daniel Norman received this plantation as a "Gift from his father-in-law." He was occupying it in 1802.

No. 756: Island Cove. Thos. Merser received this plantation as a "Gift from his father-in-law." Date is not given.

No. 896: Carbonear. Hugh Penny, Jr., received this plantation by "his father-in-law's will" in 1790.

No. 925: Crocker's Cove. Martin George received this plantation as a "gift from his father-in-law" in 1785. He was leasing it.

No. 930: Crocker's Cove. Daniel Canlan received this plantation as a "Gift from his father-in-law." Daniel was occupying it in 1785.

No. 966: Harbour Grace, North Side. Garland Bradbury received this plantation as a "Gift from his father-in-law" in 1805.

No. 972: Harbour Grace, North Side. Jas. Cawley received this plantation from "his father-in-law's will," and James was occupying it in 1715.

No. 992: Harbour Grace, North Side. Wm. Parsons received this plantation from "his father-in-law's will." Date is not given.

No. 993: Harbour Grace, North Side. Wm. Lilly, Esq., received this plantation by his "father-in-law's will." Date is not given.

BIBLIOGRAPHY

A. Books, Pamphlets and Papers

Allen, Vivien. *Lady Trader: A Biography of Mrs. Sarah Heckford.* London: Collins, 1979.

Andrews, William L. *Journeys in New Worlds: Early American Women's Narratives.* Madison: University of Wisconsin Press, 1990.

Andrieux, J. P. *Marine Disasters of Newfoundland and Labrador.* Ottawa: O. T. C. Press, 1986.

Antler, Ellen. "Women's Work in Newfoundland Fishery Families." 11-page lecture presented at a conference on women, Mount St. Vincent University, Halifax, November 13, 1976.

Armour, Moira and Pat Staton. *Canadian Women in History: A Chronology.* Toronto: Green Dragon Press, 1990.

Baird, Donal M. *Women at Sea in the Age of Sail.* Halifax: Nimbus, 2001.

Balano, James W. *The Log of the Skipper's Wife.* Camden, ME: Down East Books, 1979.

Bonnycastle, Richard Henry, Sir. *The Canadas in 1841*, Vol. 2. London: Henry Colburn, 1841. S. R. Publishers: Johnson Reprint, 1968.

-------. *Newfoundland in 1842: A Sequel to The Canadas in 1841*, Vols. 1 and 2. London: Henry Colburn, 1842.

Boorstin, Daniel J. *The Americans: The Colonial Experience.* New York: Vintage Books, 1958.

Bristol & America: A Record of the First Settlers in the Colonies of North America, 1654-1685. London: R. Sydney Glover. Centre for Newfoundland Studies, Memorial University of Newfoundland, St. John's, no publication date.

Brown, Addison. "Early History of Bell Island." Manuscript of lecture.

Centre for Newfoundland Studies, Memorial University of Newfoundland, St. John's, no date given.

Burns, Eleanor. *Egg Money Quilts: 1930s Vintage Samplers*. Sann Marcos, CA: Quilt in a Day, 2005.

Butt, Arthur S. *Telling It As It Was: A Brief History of Flat Islands and an Autobiography*. Glovertown, NL: Glovertown Literary Creations, 1990.

Cadigan, Sean T. *Economic and Social Relations of Production on the northeast coast of Newfoundland, with special reference to Conception Bay, 1785-1855*. Ph.D. Thesis, Memorial University of Newfoundland, St. John's, 1991.

-------. *Hope and Deception in Conception Bay: Merchant and Settler Relations in Newfoundland, 1785-1855*. Toronto: University of Toronto Press, 1995.

Cell, Gillian T. *English Enterprise in Newfoundland 1577-1660*. Toronto: University of Toronto Press, 1969.

Cell, Gillian T. (Ed.). *Newfoundland Discovered: English Attempts at Colonization, 1610-1630*. London: Hackluyt Society, 1982.

Cleal, Thomasina. "Women in Newfoundland and Labrador History." 14-page speech delivered at the Women's Network Dinner Meeting, April 26, 1982.

Coffin, Captain James Doane. *Journal of the* Margaret Rait, *1840-1844*. Introduction by Marion Robertson. Hantsport, NS: Lancelot Press, Paperback edition, 1984.

Coffin, Robert P. Tristram. *Captain Abby and Captain John: An Around-the-World Biography*. New York: Macmillan, 1939.

Concerning Landed Property in Newfoundland. Handwritten document, pp. 373-388. Maritime History Archive, Memorial University of Newfoundland, St. John's, 1786.

Conrad, Margaret. *Recording Angels: The Private Chronicles of Women from the Maritime Provinces of Canada, 1750-1950*. Ottawa: Canadian Research Institute for the Advancement of Women, 1982 (The CRIAW Papers, no. 4).

Conrad, Margaret, Toni Laidlaw and Donna Smyth (Eds.). *No Place Like Home: Diaries and Letters of Nova Scotia Women 1771-1938*. Halifax: Formac Publishing Company, 1988.

Conway, Jill K., with the assistance of Linda Kealey and Janet E. Schulte. *The Female Experience in Eighteenth- and Nineteenth-Century America: A Guide to the History of American Women.* Princeton: Princeton University Press, 1985.

Crevecoeur, Hector St. John de. *Letters from an American Farmer.* London: J. M. Dent, 1912.

Crosbie, Gert. *Births, Deaths, Marriages in Newfoundland Newspapers.* Maritime History Archive, Memorial University of Newfoundland, St. John's, 1986.

Davis, Dona Lee. *Blood and Nerves: An Ethnographic Focus on Menopause.* St. John's: Institute of Social and Economic Research, Memorial University of Newfoundland, 1988 (Social and Economic Studies, no. 28).

DePauw, Linda Grant. *Seafaring Women.* Boston: Houghton Mifflin, 1982.

Devine, P. K. *Ye Olde St. John's (1750-1936).* Reference Department, A. C. Hunter Public Library, St. John's, no publisher, no date.

Dictionary of Canadian Biography. Toronto: University of Toronto Press. Various volumes.

Dictionary of Newfoundland and Labrador Biography. St. John's: Harry Cuff Publications, 1990.

Dranoff, Linda S. *Women in Canadian Law.* Toronto: Fitzhenry & Whiteside, 1977.

Encyclopedia of Newfoundland and Labrador. Vols. 1-5. St. John's: Newfoundland Book Publishers (1967) Ltd., 1981-1994.

Evans, Allan. *The Splendor of St. Jacques.* St. John's: Harry Cuff Publications, 1981.

Evans, Calvin D. *For Love of a Woman: The Evans Family and a Perspective on Shipbuilding in Newfoundland.* St. John's: Harry Cuff Publications, 1992.

Feild, Edward. *Journal of the Bishop of Newfoundland's Voyage of Visitation and Discovery...in the Ship 'Hawk' 1848.* London: S. P. C. N., 1849.

Gardner, Francis William. *Registration of British Ships: A Handbook on the Law and Practice relating to the Registration of British Ships, Bills of Sale and Mortgages and Transmission of Interest by reason of Death,*

Bankruptcy, Marriage, Orders of Courts, and Other Causes. London: Spottiswoode, 1896.

Gosse, Eric Martin. *The Settling of Spaniard's Bay*. St. John's: Harry Cuff Publications, 1988.

Green, H. Gordon. *Don't Have Your Baby in a Dory*. Montreal: Harvest House, 1976.

Greenhill, Basil and Ann Giffard (commentary). *Women Under Sail: Letters and journals concerning eight women travelling or working in sailing vessels between 1829 and 1949*. Newton Abbot: David & Charles, 1970.

A Guide to the Archival Holdings of the Centre for Newfoundland Studies. 3d. ed. St. John's: Memorial University of Newfoundland, 1994.

Gunn, Gertrude E. *The Political History of Newfoundland 1832-1864*. Toronto: University of Toronto Press, 1966.

Haigh, W. O. (Bill). *Foote Prints among the Kauri*. Printed in Whangarei, New Zealand: Allied Graphics Limited, 1995 [?].

Hammond, Rev. John W. *The Beautiful Isles (a History of Bell Island from 1611-1896)*. Printed by the Pentecostal Assemblies of Newfoundland. 2d. ed. 1979.

Handcock, W. Gordon. *Soe longe as there comes no women: Origins of English Settlement in Newfoundland*. St. John's: Breakwater Books, 1989.

-------. *The Story of Trinity*. Trinity Historical Society, 1997.

Harris, Cyril. *Northern Exposure: A Nova Scotia Boyhood*. New York: W. W. Norton, 1963.

Harris, Michael. *Rare Ambition: The Crosbies of Newfoundland*. Markham, ON: Viking, 1992.

Harvey, Rev. M. *Newfoundland in 1900*. New York: South Publishing: St. John's: S. E. Garland, 1900.

-------. *A Short History of Newfoundland: England's Oldest Colony*. London: William Collins, Son & Co., 1890.

Head, C. Grant. *Eighteenth Century Newfoundland*. Toronto: McLelland & Stewart, 1976.

The History of Burin by its Senior Citizens. (Fourth Edition). Burin, NL. 1977.

Holcombe, Lee. *Wives and Property: Reform of the Married Women's Property Law in Nineteenth Century England*. Toronto: University of Toronto Press, 1983.

Hotten, John Camden. *The Original List of Persons of Quality:...children stolen, maidens pressed, and others...who went from Great Britain to the American Plantations, 1600-1700.* 2nd. ed. New York: J. W. Bouton, 1880.

Howell, Colin, and Richard J. Twomey (Eds.). *Jack Tar in History: Essays in the History of Maritime Life and Labour.* Fredericton, NB: Acadiensis Press, 1991.

Jamieson, A. G. (Ed.). *A People of the Sea: The Maritime History of the Channel Islands*. London: Methuen, 1986.

Jukes, J. B. *Excursions in and about Newfoundland during the years 1839 and 1840.* Vols. I and II. London: John Murray, 1842. Toronto: Canadiana House, 1969.

Junger, Sebastian. *The Perfect Storm: A True Story of Men Against the Sea.* New York: HarperTorch, 1997.

Kealey, Linda (Ed.). *Pursuing Equality: Historical Perspectives on Women in Newfoundland and Labrador.* St. John's: Institute of Social and Economic Research, Memorial University of Newfoundland, 1993 (Social and Economic Papers, no. 20).

Kennedy, William Robert, Sir. *Sport, travel and adventure in Newfoundland and the West Indies*. Edinburgh and London: William Blackwood and Sons, 1885.

Kirke, Henry. *The First English Conquest of Canada: With some account of the earliest settlement in Nova Scotia and Newfoundland*. 2nd. ed. Enl. London: Sampson Low, Marston, 1908.

Kirwin, William J., G. M. Story, and Patrick O'Flaherty (Eds.). *Reminiscences of James P. Howley: Selected Years*. Toronto: The Champlain Society, 1997.

Langdon-Davies, John. *A Short History of Women*. New York: Viking Press, 1927.

Lawrence, Mary Chipman. *The Captain's Best Mate: The Journal of Mary Chipman Lawrence on the whaler* Addison *1856-1860*. Hanover: University Press of New England, 1966.

Light, Beth and Alison Prentice (Eds.). *Pioneer and Gentlewomen of British North America, 1713-1867.* Toronto: New Hogtown Press, 1980.

Lounsbury, Ralph G. *The British Fishery at Newfoundland 1634-1763.* New York: Archon Books, Reprint 1969 (Yale University Press, c1934).

McCarthy, Michael J. *Women of Newfoundland 1003-1800: Their Trials, Tribulations and Triumphs.* St. John's, 1976.

McGrath, Carmelita, Barbara Neis, and Marilyn Porter (Eds.). *Their Lives and Times: Women in Newfoundland and Labrador, A Collage.* St. John's: Killick Press, 1995.

McGrath, P. T. *Newfoundland in 1911.* London: Whitehead, Morris & Co., 1911.

McGregor, J. *Historical and Descriptive Sketches of the Maritime Colonies of British America.* London: Longman, Rees, Orme, Brown and Green, 1828.

Major, Kevin. *As Near to Heaven by Sea: A History of Newfoundland and Labrador.* Toronto: Penguin Books (Canada), 2001.

Mannion, John J. (Ed.). *The Peopling of Newfoundland: Essays in Historical Geography.* St. John's: Institute of Social and Economic Research, Memorial University of Newfoundland, 1977 (Social and economic papers, no. 8).

Manuel, Edith. "Women in Newfoundland History." 21-page lecture delivered to the Newfoundland Historical Society, St. John's, February 24, 1976.

Martin, R. Montgomery. *History of Nova Scotia, Cape Breton, The Sable Islands, New Brunswick, Prince Edward Island, The Bermudas, Newfoundland, etc., etc.* London: Whittaker & Co., 1837.

Martin, Wilfred B. W., with the assistance of Eileen Martin. *Random Island Pioneers.* St. John's: Creative Publishers, 1990.

Mason, John. *A Brief Discourse on the New-Found-Land.* Edinburgh: Andro Hart, 1620.

Matthews, Keith. *Lectures on the History of Newfoundland, 1500-1830.* St. John's: Breakwater Books, 1988.

-------. *Who Was Who of Families Engaged in the Fishery and Settlement of Newfoundland, 1660-1840.* St. John's: Memorial University of Newfoundland, 1971.

Mott, Henry Youmans. *Newfoundland Men: A Collection of Biographical Sketches.* Concord, New Hampshire: T. W. & J. F. Cragg, 1894.

Moyles, R. G. *"Complaints is many and various, but the odd Divil likes it": Nineteenth Century Views of Newfoundland.* Toronto: Peter Martin Associates, 1975.

Mrs. Ricketson's Whaling Journal, 1871-1874: The Journal of Annie Holmes Ricketson on the Whaleship A. R. Tucker. New Bedford, MA: Reynolds-DeWalt, 1958.

Murray, Hilda Chaulk. *More Than Fifty Percent: Woman's Life in a Newfoundland Outport 1900-1950.* St. John's: Breakwater Books, 1979.

Murray, Jean M. (Ed.). *The Newfoundland Journal of Aaron Thomas, 1794. Able Seaman in H. M. S.* Boston. Don Mills, ON: Longmans Canada, 1968.

Nadel-Klein, Jane and Dona Lee Davis (Eds.). *To Work and To Weep: Women in Fishing Economies.* St. John's: Institute of Social and Economic Research, Memorial University of Newfoundland, 1988 (Social and Economic Papers, no. 18).

Never Done: Three Centuries of Women's Work in Canada. Toronto: Canadian Women's Educational Press, 1974.

O'Flaherty, Patrick. *Lost Country: The Rise and Fall of Newfoundland, 1843-1933.* St. John's: Long Beach Press, 2005.

-------. *Old Newfoundland: A History to 1843.* St. John's: Long Beach Press, 1999.

Ommer, Rosemary E. *From Outpost to Outport: A Structural Analysis of the Jersey-Gaspe Cod Fishery, 1767-1886.* Montreal: McGill-Queen's University Press, 1991.

Paine, Ralph D. *The Ships and Sailors of Old Salem: The Record of a Brilliant Era of American Achievement.* [Revised with a new preface by the author and with a new and complete index.] Boston: Charles E. Lauriat, 1923 (c1908).

Parker, John P. *Sails of the Maritimes: The Story of the Three- and Four-Masted Cargo Schooners of Atlantic Canada, 1859-1929.* Toronto: McGraw-Hill Ryerson, 1960.

Pedley, Rev. Charles. *The History of Newfoundland from the Earliest Times to the Year 1860.* London: Longman, Green, Longman, Roberts & Green, 1863.

The Penny Story 1856-1956. Ramea, Newfoundland. Centre for Newfoundland Studies, Memorial University of Newfoundland, St. John's, unpaged.

Pickett, Patrick (Ed.). *A History: Town of Fogo, Newfoundland, 1728-1922.* Published by the Seaside Retired Citizens Club. Printed in Grand Falls-Windsor by Robinson-Blackmore, 1992.

Pope, Peter Edward. *The South Avalon Planters, 1630 to 1700: Residence, Labour, Demand and Exchange in Seventeenth-Century Newfoundland.* Ph.D. Dissertation, Memorial University of Newfoundland, St. John's, 1992.

Porter, Marilyn. *Women and Old Boats: The Sexual Division of Labour in a Newfoundland Outport.* 25-page paper. Honorary Research Fellow: Department of Sociology, University of Manchester, 1982. Published in Gender and Society. London: Heinemann, 1983.

Prowse, Daniel W. *A History of Newfoundland from the English, Colonial and Foreign Records.* (2d. ed. rev. and corr.) London: Eyre and Spottiswoode, 1896.

Pumphrey, Ron. *Who's Who and Why – on the "Boot of the Burin" and in St. John's.* Printed by Robinson-Blackmore, 1985.

Rae, W. Fraser. *Newfoundland to Manitoba.* London: Sampson Low, Marston, Searle & Rivington, 1881.

"Remarkable Women of Newfoundland and Labrador." Presented by the St. John's Local Council of Women. Valhalla Press Canada, 1976. Centre for Newfoundland Studies, Memorial University of Newfoundland, St. John's.

Rothney, G. O. *Newfoundland: A History.* 2d. ed. Ottawa: Canadian Historical Association, 1964, c 1959 (Historical booklet, no. 10).

Sager, Eric W. and Lewis R. Fischer. *Shipping and Shipbuilding in Atlantic Canada, 1820-1914.* Ottawa: Canadian Historical Association, 1986 (Historical booklet, no. 42).

St. Peter's Anglican Church, Twillingate. *One Hundred and Twenty-five Year History, 1845-1970.* Printed by the Morgan Printing Co., 1970.

Salmon, Marylynn. *Women and the Law of Property in Early America.* Chapel Hill, NC: University of North Carolina Press, 1986.

Seary, E. R. *Family Names of the Island of Newfoundland.* St. John's: Memorial University of Newfoundland, 1977.

Shanley, Mary Lyndon. *Feminism, Marriage, and the Law in Victorian England, 1850-1895.* Princeton: Princeton University Press, 1989.

Smallwood, Joseph R. (Ed.). *Book of Newfoundland,* Vols. 1-2, 1937. Vols. 3-4, 1967. St. John's: Newfoundland Book Publishers.

Smith, Richard M. (Ed.). *Land, Kinship and Life-Cycle.* Cambridge: Cambridge University Press, 1984.

Snow, Alice Rowe. *Log of a Sea Captain's Daughter with adventures on Robinson Crusoe's Island, by Alice Rowe Snow, who sailed with her father, Captain Joshua N. Rowe, on the Bark* Russell *during a voyage of four years and a half.* Boston: Meador Press, 1944.

Snow, Edward Rowe. *Women of the Sea.* New York: Dodd, Mead, 1962. (New ed. 2004).

Spicer, Stanley T. *Masters of Sail: The Era of Square-Rigged Vessels in the Maritime Provinces.* Toronto: McGraw-Hill Ryerson, 1968.

Story, G. M., W. J. Kirwin, and J. D. A. Widdowson. (Eds.) *Dictionary of Newfoundland English.* Toronto: University of Toronto Press, 1982.

Those Bright Days…Come Home Year 1992…Boxey, Wreck Cove, English Hr. West, Mose Ambrose, St. Jacques, Coombs Cove. Reference Department, A. C. Hunter Public Library, St. John's.

Tocque, Rev. Philip. *Newfoundland: As it was, and as it is in 1877.* Toronto: John B. Magurn, 1878.

Tute, Warren. *Atlantic Conquest: The Men and Ships of the Glorious Age of Steam.* Boston: Little, Brown & Co., 1962.

Uttley, John. *The Story of the Channel Islands.* London: Faber & Faber, 1966.

Villiers, Alan John. *Men, Ships and the Sea.* Washington: National Geographic Society, 1973.

Wallace, Frederick William. *Wooden Ships and Iron Men: The Story of the Square-Rigged Merchant Marine of British North America.* London: Hodder and Stoughton, 1924.

Walsh, Alice. *Heroes of Isle aux Morts.* Toronto: Tundra Books, 2001.

Whiffen, Bruce. *Prime Berth: An Account of Bonavista's Early Years.* St. John's: Harry Cuff Publications, 1993.

Whiteley; William H. *Duckworth's Newfoundland: The Island in the Early Nineteenth Century*. St. John's: Harry Cuff Publications, 1985.

-------. *James Cook in Newfoundland, 1762-1767*. St. John's: Newfoundland Historical Society, 1975. (Pamphlet Number 3).

Whiting, Emma Mayhew and Henry Beetle Hough. *Whaling Wives*. Boston: Houghton Mifflin, 1953.

Wilkie, D. *Sketches of a Summer Trip to New York and the Canadas*. Edinburgh: Sherwood, Gilbert, and Piper. London: J. Anderson, Jun., and A. Hill, 1837.

Williams, Harold (Ed.). *One Whaling Family*. Boston: Houghton Mifflin, 1964.

Winsor, Fred. *The Newfoundland Bank Fishery: Government Policies and the Struggle to Improve Bank Fishing Crews' Working, Health, and Safety Conditions, 1876-1920*. Ph.D. Thesis, Memorial University of Newfoundland, St. John's, 1996.

Wix, Edward. *Six Months of a Newfoundland Missionary's Journal, from February to August 1835*. London: Smith, Elder, 1836.

Woloch, Nancy. *Early American Women: A Documentary History, 1600-1900*. Belmont, CA: Wadsworth, 1992.

Women and the Law in Newfoundland. 1st ed. published by St. John's Women's Bureau, 1972. 2d. ed. rev. by Lillian Bouzane. St. John's: Newfoundland Status of Women Council, 1976.

Women and the Law in Newfoundland and Labrador. St. John's: Provincial Advisory Council on the Status of Women, 1987.

B. Periodical Articles

Bonham, Julia C. "Feminist and Victorian: The Paradox of the American Seafaring Woman of the Nineteenth Century." *American Neptune*, Vol. 37, 1977, pp. 203-218.

Bowen, Noel H. "The Social Condition of the Coast of Labrador." In *Report of the Literary and Historical Society of Quebec*, 1855, 334. Excerpt in *Pioneer and Gentlewomen of British North America*, 1713-1867, p. 159.

Christensen, Ruth M. "The Establishment of S. P. G. Missions in Newfoundland, 1703-1783." *The Historical Magazine of the Protestant Episcopal Church*, Vol. 20, 1951, pp. 207-229.

Dugaw, Dianne. "'Rambling Female Sailors': The Rise and Fall of the Seafaring Heroine." *International Journal of Maritime History*, Vol. 4, no. 1, June 1992, pp. 179-194.

Greer, Germaine. "All her worldly goods were his." Germaine Greer on what English women in the early modern age could own – whether a "landed estate," a cow or a kettle. Review of Amy Louise Erickson's book, *Women and Property in Early Modern England. The Guardian*, April 30, 1994.

Handcock, W. G. "Spatial Patterns in a Trans-Atlantic Migration Field: The British Isles and Newfoundland During the Eighteenth and Nineteenth Centuries." *Proceedings of the 1975 British-Canadian Symposium on Historical Geography. The Settlement of Canada: Origins and Transfer*, pp. 13-45. Kingston, ON: Queen's University, 1976.

Harcourt, Freda, Simon Ville, and one other. "Notes on Sarah Palmer's *Politics, Shipping and the Repeal of the Navigation Laws* (1990) with a Response by Sarah Palmer." Roundtable Commentary. *International Journal of Maritime History*, 1991, pp. 227 to 237ff.

Maloney, Aiden. "King's Cove." *The Newfoundland Quarterly*, Vol. 88, no. 3, April 1994, pp. 15-20.

Morris, Don. "U.S. President and the Widow Huelen." In column *Vignettes of the West* in the *Western Star*. Original date unknown. Transcribed by Steve Gillis with permission granted by the *Western Star* and posted to the Internet 2001.

Tucker, Ephraim W. "Five Months in Labrador and Newfoundland during the Summer of 1838 (Concord, 1839)," 119-20. Excerpt in *Pioneer and Gentlewomen of British North America*, 1713-1867, p. 158-159.

Whiteley, William H. "James Cook and British Policy in the Newfoundland Fisheries," *The Canadian Historical Review*, Vol. 54, no. 3, 1973, pp. 245-272.

C. Newspapers and Magazines

Advertiser, Grand Falls-Windsor, NL.

Chronicle-Herald, Halifax, NS.

Daily News, St. John's.

Decks Awash, Botwood Historical Society Archives, NL.

Evening Telegram, St. John's.

Free Press, St. John's.

Newfoundland Ancestor. St. John's: Quarterly Journal of the Newfoundland and Labrador Genealogical Society.

Newfoundland Journal of Commerce, Reference Department, A. C. Hunter Public Library, St. John's

Newfoundland Quarterly, St. John's.

Newfoundlander, St. John's.

Public Ledger, St. John's.

Royal Gazette, St. John's.

Standard and Conception Bay Advertiser, Harbour Grace, NL.

Times and General Commercial Gazette, St. John's.

Twillingate Sun, Twillingate, NL.

Western Star, Corner Brook, NL.

D. Old Census Data for Newfoundland, by date

Sir John Berry Census, 1675. *The Names of the English Inhabitants with the number of their boats, men, wives and children (from Bonavista to Trepassy)*. Provincial Archives of Newfoundland and Labrador, St. John's. Sir John Berry did others in 1677, 1681, 1684, 1698 (heads of households only) and 1708. Only the 1675 and 1677 records were checked in this case.

Shortis, H. F. *All about Pirates. Pirates and Census, 1675*. Provincial Archives of Newfoundland and Labrador, St. John's. (This includes a census of Newfoundland from Cape Race to Cape Bonavista, 1675).

Planters living in St. John's, 1681. Provincial Archives of Newfoundland and Labrador, St. John's.

Smith, Warwick. *Trinity 1753-1768.* Typed manuscript. Signed 15 December 1939. Provincial Archives of Newfoundland and Labrador, St. John's.

Handcock, W. Gordon. *Historical Documents and Reference Material on Trinity, Census 1753.* Maritime History Archive, Memorial University of Newfoundland, St. John's.

Old Properties. Early Residents of St. John's, 1780-1781. A Project of the Newfoundland Archives. Authorized by the Minister of Provincial Affairs. Provincial Archives of Newfoundland and Labrador, St. John's.

Census of St. John's for 1794-1795. Colonial Secretary's Office. Provincial Archives of Newfoundland and Labrador, St. John's.

Kirwin, William and James Feltham. *An Index to the Names in an Account of Inhabitants residing in the Harbour and Districts of St. John's, 1794/1795.* Public Reference Library, Department of English Language and Literature, Memorial University of Newfoundland, St. John's, 1968.

Shortis, H. F. *Old St. John's. Partial Census 1796-1797.* Written by H. F. Shortis, Esquire. Reprinted from *The Cadet*, 1916. A project of the Newfoundland & Labrador Provincial Archives authorized by Hon. J. R. Chalker, Minister of Provincial Affairs. Provincial Archives of Newfoundland and Labrador, St. John's, printed in 1971.

Return of the Number of Houses, Inhabitants, fishing rooms...in the winter of 1800 and 1801, with the number of servants and total of wages given (for) Trinity. Provincial Archives of Newfoundland and Labrador, St. John's.

Number of Inhabitants in the Harbours of Brigus, Cupids, Bareneed, Port de Grave for 1817. Provincial Archives of Newfoundland and Labrador, St. John's.

Payne, Brian M. *An Analysis of Criminal and Civil Cases in Ferryland and Placentia Court Records.* Paper, 1972. Maritime History Archive, Memorial University of Newfoundland, St. John's.

E. Registers of Fishing Rooms

Centre for Newfoundland Studies, Memorial University of Newfoundland, St. John's.

Conception Bay Plantation Book 1805, including "Back Rooms." 158 pages; 1062 entries; about 147 women referenced in the Plantation Book.

Survey of Plantations in Conception Bay 1807. (Largely a duplication of previous entry).

Register of Fishing Rooms in Bonavista 1805-06. Total of 91 Rooms. At least seven women referenced.

Twillingate and Places Adjacent. Register of Fishing Rooms, January 3, 1806. Total of 29 Rooms. At least six women referenced.

F. Miscellaneous

An Act for declaring all landed property in Newfoundland real chattels, June 12, 1834. Legislative Library, House of Assembly, St. John's.

The Consolidated Statutes of Newfoundland, 1871. Chapter XCV. Of the Protection of Married Women. Centre for Newfoundland Studies, Memorial University of Newfoundland, St. John's.

An Act to Amend the Law Relating to the Property of Married Women, April 26, 1876. Legislative Library, House of Assembly, St. John's.

An Act to Amend the Law Relating to the Property of Married Women, April 21, 1883. Legislative Library, House of Assembly, St. John's.

An Act to Amend the Married Women's Property Act of 1883, passed July 4, 1895. Legislative Library, House of Assembly, St. John's.

The Consolidated Statutes of Newfoundland, 1916. Chapter 112. Of the Property of Married Women. Centre for Newfoundland Studies, Memorial University of Newfoundland, St. John's.

Acts of the General Assembly of Prince Edward Island, 1903. Charlottetown: John Coombs, King's Printer, 1903. Government Documents Department, McGill University Libraries, Montreal.

Acts of the General Assembly of Her Majesty's Province of New Brunswick, passed in the months of February and March, 1895. Fredericton: G. E. Fenety, Printer to the Queen's Most Excellent Majesty, 1895. Government Documents Department, McGill University Libraries, Montreal.

Revised Statutes of Nova Scotia, 1900. Government Documents Department, McGill University Libraries, Montreal.

Quebec. Civil Code. 1866. Chapter Third. II. Of the clause of separation of property. Government Documents Department, McGill University Libraries, Montreal.

Statuts de Quebec, 1963-64. Government Documents Department, McGill University Libraries, Montreal.

Quebec. Registrar of Shipping. Scrapbook of directives to the Registrar of Shipping from Dept. of Marine & Fisheries, and registers, letterbooks, etc. pertaining to shipping in a particular port, 1850-1958. Ontario Public Archives, Ottawa.

Ferryland Petition. 1696, November 27. To his most sacred Majesty, William the 3rd of England & Scotland, France and Ireland, King, Defender of the Faith. (re. French aggression). Provincial Reference Department, A.C. Hunter Library, St. John's.

Court Records. Placentia, 1757-1786, 1786-1803, 1805-1806. Provincial Reference Department, A.C. Hunter Library, St. John's.

-------. St. John's, 1758-1759. Provincial Reference Department, A.C. Hunter Library, St. John's.

-------. Ferryland, 1773-1786. Provincial Reference Department, A.C. Hunter Library, St. John's.

-------. Trepassey, 1789-1804. Provincial Reference Department, A.C. Hunter Library, St. John's.

-------. Burin, 1825. Provincial Reference Department, A.C. Hunter Library, St. John's.

Personal Names File. Provincial Reference Department, A.C. Hunter Public Library, St. John's.

G. Manuscript Collections, by date
All in Provincial Archives of Newfoundland and Labrador, St. John's.

Carter-Barrett Bill of Sale, 1711, Old Perlican.

Burt Family Papers, Old Perlican, early 1700s.

John Slade & Co., Twillingate, c. 1783-1862.

Parsons Family, Harbour Grace, c. 1792-1959.

Ferryland Papers – Carter and Nunan, 1810-1850.

Morris & Ryan, merchants, Burin and St. John's, c. 1811-1826.

Thomas Slade and William Cox, Twillingate, c. 1815-1840.

Richard Morris of St. John's, c. 1817-1826.

Nathaniel Woodley Collection, St. John's, c. 1825-1866.

Bolton & Bolton, Carbonear. Ledger and Daybooks, 1834-37.

Fowlow Family of Trinity, c. 1837-1848. Reference to Richard and Amelia Purchase, 1837.

Mary Wells Collection, English Harbour, c. 1839-1876.

Mary A. Simms Collection, Ferryland, c. 1839-1888.

Marshall Family of Burin, c. 1857-1888.

Cape Broyle & Area, (Rough Diary), 1914-1946.

H. Letters, Pictures and Completed Questionnaires received from:

Ruth Beckley, Winchester, Massachusetts

Christie Bradley, Grand Bank, Newfoundland

Oliver K. Black, Salem, Massachusetts

James A. H. Carr, Toronto (and Grand Bank, Newfoundland)

Carrie R. Dancey, St. John's (and Burin, Newfoundland)

Ann Eddy, North Harbour, Placentia Bay, Newfoundland

Tom Foote, Dartmouth, Nova Scotia (and Grand Bank, Newfoundland)

Eric Martin Gosse, Spaniard's Bay, Newfoundland

Evelyn Grondin, Curator, Burin Heritage House, Newfoundland

Gerald Hann, Petite Forte, Newfoundland

Eliza Hearn, St. John's (and Harbour Breton, Newfoundland)

Earle Hickey, Petite Forte, Newfoundland

Lloyd and Malvina Hollett, Burin Bay Arm, Newfoundland

Wayne Hollett, Burin, Newfoundland

Eva Linehan, Branch, Newfoundland

Stephen J. March, Old Perlican, Newfoundland

Margaret A. O'Brien, Cape Broyle, Newfoundland

Debbie Petite, English Harbour West (and Mose Ambrose, Newfoundland)

Maureen Regan, Halifax, Nova Scotia (and Cape Broyle, Newfoundland)

Dr. Grace M. Sparkes, St. John's (and Grand Bank, Newfoundland)

Capt. Harry W. Stone, St. John's (and Monroe, Trinity Bay, Newfoundland)

Beverly Warford, Point Leamington, Newfoundland

INDEX

Women and Ships

ACKNOWLEDGEMENTS

My thanks are due to many people and several institutions. I am grateful to McGill University for awarding me a full-year sabbatical leave to undertake the research for this book and to the Institute of Social and Economic Research (ISER) at Memorial University of Newfoundland for awarding me a research grant to visit about 20 communities in Newfoundland.

I am grateful to Heather Wareham, Archivist at the Maritime History Archive at Memorial University, and to all her staff for their invaluable assistance, and to Heather, in particular, for arranging for me to give a lecture to Maritime historians and other interested parties in 1994. Thanks go as well to: Professors Valerie Burton and Daniel Vickers of the Maritime Studies Research Unit and the Department of History at Memorial University for their support and encouragement; staff at the Centre for Newfoundland Studies and their archives at Memorial University; staff in the Reference Department at the Newfoundland Public Libraries at the St. John's Arts & Culture Centre; staff at the Newfoundland Provincial Archives; staff at the McGill University Libraries; staff at the National Archives of Canada in Ottawa; curators and other staff at various museums in Newfoundland; Betty Williams at the Registrar of Shipping in St. John's; Trudi Johnson, who was in 1994 a Ph.D. student at Memorial University and who provided much valuable information on property ownership by women.

Additionally, I would like to thank: Professor Gordon Handcock, Professor Chris English, and Professor Lewis R. Fischer at Memorial University for valuable advice and encouragement; Professor Carmen Miller at McGill University for inviting me to lecture on this subject to his history class; McGill University Libraries for sponsoring a public lecture on the topic in 1995; Clifford Evans, Curator of the Mary March Museum at Grand Falls-Windsor, for arranging a public lecture in the museum in 1994; retired Professor Otto Tucker of Memorial University for inviting me to lecture at the Wessex Society in St. John's in February 2005; and finally, staff at the public libraries in Botwood, Gander and Grand Falls-Windsor for facilitating access to research and archival information for this project.

I am especially grateful to the many people who welcomed me unannounced into their homes on the south-west coast, the south coast, the Burin Peninsula, Glovertown and the east coast of Newfoundland. I suppose there is no other place in all the world where one can be so readily accepted and warmly received and never made to feel that one is intruding. My wife accompanied me on most of the visits, and on those occasions when she waited in the car, reading a book, she was routinely invited into a nearby house for a chat and a cup of tea. I was looking only for information but I came away feeling that total strangers had become friends.

I supplemented these visits by sending a standardized questionnaire to other Newfoundland communities simply by selecting from the telephone directory a person by the same name as the woman shipowner. I was especially pleased with the enthusiastic and helpful responses. Some even followed up those initial responses with new and additional information and pictures.

As a final note, I would like to thank my wife Goldie for her patience and long suffering and strong support throughout this project. She has done without me for many long hours through many months and now will be able to stop asking the question: "Are you finished yet?" I appreciate her endurance.

To all those who helped in any way, my heartfelt gratitude.

CALVIN D. EVANS retired from dual careers as a university librarian and a United Church minister. He had a 30-year career as a library administrator at Memorial University of Newfoundland, University of Guelph in Ontario, University of Alberta in Edmonton, and McGill University in Montreal. In 2005 he celebrated his 50th year of ordination and since 1949 has served churches in Newfoundland, Nova Scotia, Saskatchewan, Alberta, Ontario and Quebec. Naturally, the two careers overlapped.

He is author of two books: *For Love of a Woman: The Evans Family and a Perspective on Shipbuilding in Newfoundland*, published in 1992 by Harry Cuff Publications, St. John's, and *Soren Kierkegaard Bibliographies*, published in 1993 by McGill University. He has also written several articles and has been a contributing writer to seven volumes of the *Dictionary of Canadian Biography* and to the *Encyclopedia of Newfoundland and Labrador*. In 2005 he wrote the biographies of fifteen Newfoundland women for the Grand Falls-Windsor chapter of the Council on the Status of Women.

Calvin has an avid interest in Newfoundland history and is completing the book *Women of the Maritime Provinces and Quebec and Their Ships* and is working on another book *Virtual Communities: Early Sawmills in Northeastern and Central Newfoundland*. He was a member of the Botwood Heritage Society for ten years, and is now living with his wife Goldie at Wasaga Beach, Ontario.